Bacterial Respiration and Photosynthesis

Series editors
Dr J A Cole, University of Birmingham
Dr C J Knowles, University of Kent

Titles in series
1 Oral Microbiology *P Marsh*
2 Bacterial Toxins *J Stephen and R A Pietrowski*
3 The Microbial Cell Cycle *C Edwards*
4 Bacterial Plasmids *K Hardy*
5 Bacterial Respiration and Photosynthesis *C W Jones*
6 Bacterial Cell Structure *H J Rogers*
7 Microbial Control of Pests and Diseases *J Deacon*

JONES, COLIN WILLIAM
BACTERIAL RESPIRATION AND PHOT
000491352

QR89.J71

THE UNIVERSITY OF LIVERPOOL

HAROLD COHEN LIBRARY

Please return or renew, on or before the last date below. A fine is payable on late returned items. Books may be recalled after one week for the use of another reader. Unless overdue, or during Annual Recall, books may be renewed by telephone:- 794 - 5412.

DUE TO RETURN

29 JUN 1996

For conditions of borrowing, see Library Regulations

Aspects of Microbiology 5

Bacterial Respiration and Photosynthesis

Colin W Jones
Senior Lecturer, Department of Biochemistry, University of Leicester

Nelson

Acknowledgements

I am indebted to Jan Drozd, Hilary Evans, Bruce Haddock, Allan Hamilton, Barry Jackson and Don Kelly for their comments on individual chapters or passages, and to Michael Dawson and Alan McKay for criticising and proof-reading the entire manuscript. My thanks are also due to Jeff Cole for his particularly helpful and conscientious editorial work, and to Amelia Dunning for efficiently typing parts of the manuscript. Finally I would like to express my gratitude to my wife, Beryl, both for her unstinting secretarial work and for her encouragement during the writing of this book.

C W JONES

Thomas Nelson and Sons Ltd
Nelson House Mayfield Road
Walton-on-Thames Surrey KT12 5PL

P O Box 18123 Nairobi Kenya

116-D JTC Factory Building
Lorong 3 Geylang Square Singapore 1438

Thomas Nelson Australia Pty Ltd
19–39 Jeffcott Street West Melbourne Victoria 3003

Nelson Canada Ltd
81 Curlew Drive Don Mills Ontario M3A 2R1

Thomas Nelson (Hong Kong) Ltd
Watson Estate Block A 13 Floor
Watson Road Causeway Bay Hong Kong

Thomas Nelson (Nigeria) Ltd
8 Ilupeju Bypass PMB 21303 Ikeja Lagos

© Colin W Jones 1982
First published 1982

ISBN 0-17-771105-1
NCN 420-5815-0

All rights reserved. No part of this publication may be reproduced,
stored in a retrieval system, or transmitted, in any form or by any means,
electronic, mechanical, photocopying, recording or otherwise,
without the prior permission of the publishers.

Photosetting by Thomson Press (India) Ltd., New Delhi
Printed and bound in Hong Kong

Contents

1 Introduction   1
   Respiration   1
   Photosynthesis   4
   Respiratory chain and photosynthetic phosphorylation   6
   The coupling membrane   10
   Summary   12
   References   13

2 Aerobic respiration in chemoheterotrophs and facultative phototrophs   14
   Respiratory chain composition   14
   Pathways of respiration   22
   The control of respiration   24
   The spatial organization of the respiratory chain   25
   Respiration-linked proton translocation   26
   Energy coupling sites   31
   The protonmotive force (Δp)   32
   Summary   36
   References   36

3 Aerobic respiration in chemolithotrophs; anaerobic respiration   38
   The oxidation and reduction of nitrogen compounds   38
   The oxidation and reduction of sulphur compounds   50
   The oxidation of hydrogen   56
   The oxidation of ferrous iron ($Fe^{2+} \to Fe^{3+}$)   57
   The reduction of ferric iron ($Fe^{3+} \to Fe^{2+}$)   58
   The reduction of fumarate to succinate   58
   The reduction of carbon dioxide to methane   59
   The reduction of trimethylamine-N-oxide to trimethylamine   61
   The oxidation of carbon monoxide to carbon dioxide   61
   Summary   61
   References   62

4 Photosynthesis   64
   Bacteriochlorophyll-dependent photosynthesis   64
   Chlorophyll-dependent photosynthesis   78
   Bacteriorhodopsin-dependent photosynthesis   81
   Summary   84
   References   85

Contents

5 Energy transduction	**86**
The ATP phosphohydrolase ($BF_0.BF_1$)	86
$BF_0.BF_1$: ATPase or ATP synthetase?	90
Energy transduction mutants	91
The energetics of ATP synthesis and hydrolysis	95
The mechanism of ATP synthesis	98
Further aspects of membrane energy transduction	100
Summary	104
References	104
Index	**105**

1 Introduction

Energy conservation in bacteria, as in higher organisms, occurs principally via the synthesis of adenosine-5′-triphosphate (ATP) from adenosine-5′-diphosphate (ADP) and inorganic phosphate:

$$H^+ + ADP^{3-} + HPO_4^{2-} \rightleftharpoons ATP^{4-} + H_2O$$

Since the hydrolysis of ATP under standard conditions releases a moderate amount of free energy ($\Delta G^{\theta'} = -31.0$ kJ.mole^{-1}), its synthesis requires a similar amount of energy ($\Delta G^{\theta'} = +31.0$ kJ.mole^{-1}), the latter being provided by specific metabolic reactions within the cell.

Bacteria use three quite distinct methods for synthesizing ATP: substrate-level phosphorylation, respiratory chain (or oxidative) phosphorylation and photosynthetic (or photo) phosphorylation. Each method involves one or more oxidation-reduction (redox) reactions, but the way in which these exergonic reactions are coupled to the endergonic condensation of ADP and phosphate is fundamentally different in substrate-level phosphorylation compared with the other two processes.

During a redox reaction, reducing equivalents (H, H$^-$ or e$^-$) are spontaneously transferred from a compound which has a tendency to donate them (the reducing half of a low redox potential couple) to a compound which has a tendency to accept them (the oxidizing half of a higher redox potential couple) e.g.

$$DH_2 \diagdown \diagup A$$
$$D \diagup \diagdown AH_2$$

According to equilibrium thermodynamics, the amount of free energy released by this reaction under standard conditions is determined by the difference between the redox potential (E'_θ) of the donor couple (D/DH$_2$) and the acceptor couple (A/AH$_2$), according to the equation:

$$\Delta G^{\theta'} = -n.F.\Delta E'_\theta$$

where n is the number of electrons transferred, F is the Faraday constant (96.6 kJ.volt^{-1}.equiv^{-1}) and $\Delta E'_\theta$ is the difference in standard redox potential ($E'_{\theta_{ox}} - E'_{\theta_{red}}$; V or mV). An exergonic redox reaction of this type can therefore be coupled to the performance of useful work, such as the formation of an energy rich compound or the generation of a membrane-associated concentration or charge gradient, both of which can subsequently be used to drive ATP synthesis.

Thus some types of substrate level phosphorylation entail the oxidation of an organic substrate (e.g. pyruvate or 3-phosphoglyceraldehyde) by an appropriate endogenous oxidant such as NAD$^+$ to generate a non-phosphorylated intermediate with a high free energy of hydrolysis. This subsequently undergoes phosphate substitution to yield an energy-rich acyl phosphate (such as acetyl phosphate or 1,3-bis-phosphoglycerate; $\Delta G^{\theta'} \geq -43.9$ kJ.mole^{-1}) which finally

Bacterial Respiration and Photosynthesis

donates a phosphoryl group ($-PO_3^{2-}$) to ADP to form ATP. Substrate-level phosphorylation is thus a scalar (spatially-directionless) series of reactions in which chemical group transfer is catalyzed by essentially soluble cytoplasmic enzymes via the sequential stoichiometric formation of covalent intermediates. In contrast, oxidative and photosynthetic phosphorylation are membrane-bond vectorial (spatially-oriented) processes which occur via a series of sequential oxidation-reduction reactions (respiration and photosynthetic electron transfer) that involve several types of spatially-organized redox carriers (the respiratory chain and the photosynthetic electron transfer system). No covalent, energy-rich intermediates have been detected during either of these processes, and energy transfer between the redox systems and the enzyme complex which is responsible for ATP synthesis (the ATP phosphohydrolase or ATPase-ATP synthetase) appears to be effected via energized protons, i.e. an electrochemical gradient of H^+. The latter can also drive other energy-dependent membrane reactions such as reversed electron transfer, cell motility and some forms of solute transport.

Substrate-level phosphorylation is the only method of ATP synthesis that is available to a few obligately anaerobic chemoheterotrophs and, under certain oxygen-deficient growth conditions, to some facultative anaerobes. In both cases it is associated with a metabolic process known as fermentation in which reducing equivalents conserved during catabolism as NADH or as other reduced cofactors are ultimately transferred to one or more endogenous organic oxidants. Since the amount of free energy liberated during fermentation is usually small, the ATP yield during substrate-level phosphorylation is correspondingly low (e.g. the homolactic fermentation of one molecule of glucose to two molecules of lactate by members of the genus *Streptococcus* yields only two molecules of ATP net, and some closely-related fermentations yield only one). In contrast, substrate-level phosphorylation is responsible for only a small fraction of the total ATP which is synthesized by bacteria which carry out respiration (most chemoheterotrophs, all chemolithotrophs and some facultative phototrophs) or photosynthetic electron transfer (most phototrophs).

Respiration

Aerobic respiration in chemoheterotrophs and facultative phototrophs is characterized by the transfer of reducing equivalents from an organic donor, principally NADH (but also other reductants including succinate, lactate and methanol), to molecular oxygen. Since NADH oxidation occurs over a large redox potential span, ($E'_\theta NAD^+/NADH + H^+ = -320mV, \frac{1}{2}O_2/H_2O = +820mV; \Delta E'_\theta = +1140mV$), both the free energy change and the ATP yield are high (e.g. the complete oxidation of one molecule of glucose to carbon dioxide and water probably yields up to 38 molecules of ATP net in some organisms). These aerobic respiratory chains contain diverse redox carriers which include flavoproteins (Fp), iron-sulphur proteins (Fe-S), quinones and cytochromes (iron-containing haemoproteins); the first two usually form the dehydrogenases which catalyse the initial oxidation of the donor, whereas specialized autoxidizable cytochromes comprise the one or more oxidases which catalyse the terminal reduction of oxygen to water (Fig. 1.1a).

Many facultatively anaerobic and obligately anaerobic chemoheterotrophs replace oxygen with alternative acceptors, and hence catalyse anaerobic respiration. These acceptors include various oxy-anions of nitrogen and sulphur,

Introduction

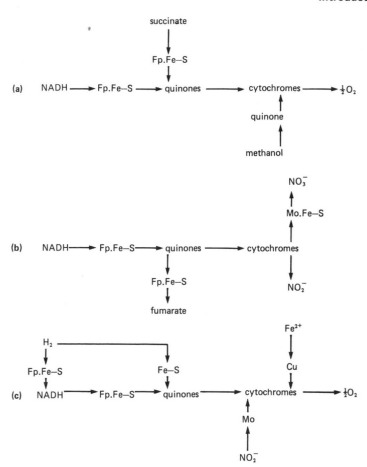

Fig. 1.1 Some examples of respiration in (**a**) aerobic chemoheterotrophs and facultative phototrophs, (**b**) anaerobic chemoheterotrophs, and (**c**) chemolithotrophs.

Fe^{3+} and organic compounds such as fumarate, carbon dioxide and trimethylamine-N-oxide; their reduction is said to be dissimilatory since the products are ultimately released into the environment. The redox potentials of these acceptors are very wide-ranging (e.g. $E'_\theta SO_3^{2-}/S^{2-} = -116\,mV$, $N_2O/N_2 = +1355\,mV$) and hence the ATP yield from anaerobic respiration is extremely varied. However, except for the reduction of Fe^{3+} and some nitrogen compounds, the yield is generally much lower than from aerobic respiration, although higher than from fermentation. Anaerobic respiratory chains contain the same types of redox carriers as those present in aerobic systems, except that the cytochrome oxidases are replaced by appropriate reductases (Fig. 1.1b). Some of the latter are novel redox carriers (e.g. molybdo-proteins and copper proteins), but the majority are specialized flavoproteins, iron-sulphur proteins or cytochromes.

Bacterial Respiration and Photosynthesis

Chemolithotrophs (chemoautotrophs) principally oxidize inorganic donors using mainly oxygen, but occasionally nitrate, as the terminal acceptor; the donors include hydrogen, various nitrogen and sulphur compounds, Fe^{2+} and, paradoxically, carbon monoxide. Since the redox potentials of these donors cover a very wide range (e.g. $E'_\theta\ CO_2/CO = -540\,mV$, $2H^+/H_2 = -420\,mV$, $Fe^{3+}/Fe^{2+} = +780\,mV$), the free energy changes and the ATP yields associated with this type of respiration also vary tremendously; however, except for the oxidation of hydrogen and probably also carbon monoxide, they are generally fairly low. Specially adapted enzymes, many of which resemble the corresponding anaerobic reductases, catalyze the initial oxidation of the inorganic donors (Fig 1c).

Each of these three types of respiration is characterized by the transfer of reducing equivalents in the direction of increasingly positive redox potential, with the concomitant release of free energy, (it should be noted that under non-standard conditions E'_θ is replaced by E_h, the actual redox potential, and that $\Delta G^{\theta'}$ is replaced by ΔG, the actual free energy change; under certain circumstances the values of E_h and ΔG may differ significantly from those of E'_θ and $\Delta G^{\theta'}$ respectively). It is possible for respiratory chains to transfer energy in the opposite direction provided that energy is put into the system (reversed respiration or reversed electron transfer). This phenomenon is particularly crucial to chemolithotrophs since these organisms require NAD(P)H for carbon dioxide assimilation. Reversed electron transfer from higher redox potential inorganic donors to $NAD(P)^+$ is driven by the proton or charge gradient generated during forward electron transfer from the same donors to oxygen or nitrate.

Photosynthesis

The overall reaction of photosynthesis in photoautotrophs may be described by the equation:

$$2H_2A + CO_2 \xrightarrow{\text{light energy}} (CH_2O) + H_2O + 2A$$

In blue-green bacteria, as well as in algae and green plants, H_2A is water and the reductive assimilation of carbon dioxide is accompanied by the release of molecular oxygen (oxygenic photosynthesis). However, many species of purple and green bacteria replace water with other inorganic reductants (e.g. H_2, S^{2-}, $S_2O_3^{2-}$), and others, photoheterotrophs, replace both water and carbon dioxide with partially reduced carbon compounds such as succinate or malate; oxygen is not released by any of these organisms (anoxygenic photosynthesis).

Two types of photosynthetic electron transfer and phosphorylation, cyclic and non-cyclic, are found in the majority of phototrophs (Fig. 1.2). The former is independent of exogenous reductants or oxidants and its sole function is to conserve light energy as a proton/charge gradient and hence as ATP (some species of purple bacteria will also synthesize inorganic pyrophosphate, $\Delta G^{\theta'} = +21.9\ kJ.mol^{-1}$, from two molecules of orthophosphate). In the first stage of this process, electro-magnetic radiation is absorbed by various specialized photopigments which include (bacterio) chlorophylls, (bacterio) pheophytins, carotenoids and, in the blue-green bacteria, various phycobiliproteins. This leads, via a series of complex photochemical reactions, to the generation of a low redox potential reductant (a reduced iron-sulphur protein in green and blue-green bacteria, and a novel quinone-iron complex in purple bacteria; $E'_\theta \leq -160\,mV$) and

Introduction

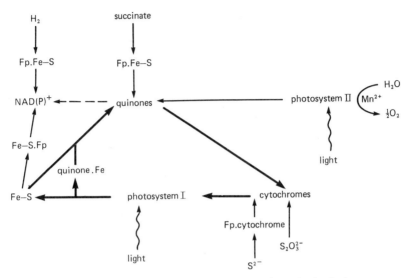

Fig. 1.2 Photosynthetic electron transfer. Cyclic electron transfer (→), the additional steps of non-cyclic electron transfer (→), reactions driven by the proton/charge gradient (-→). See text for further explanations.

a high redox potential oxidant (oxidized (bacterio) chlorophyll; $E'_\theta = +250$ mV). The transfer of reducing equivalents from the former to the latter occurs via a conventional quinone-cytochrome system and leads to the release of a medium amount of free energy and hence to a moderate ATP yield. In contrast, the principal function of non-cyclic electron transfer is to reduce $NAD(P)^+$ using an exogenous donor, the resultant NAD(P)H being used for the assimilation of carbon dioxide and for various other metabolic reactions. Since the exogenous reducing couple, with the exception of $2H^+/H_2$, has a higher redox potential than the $NAD(P)^+/NAD(P)H$ couple (e.g. $E'_\theta S_0/S^{2-} = -99$ mV, fumarate/succinate $= +30$ mV, $\frac{1}{2}O_2/H_2O = +820$ mV), the reduction of $NAD(P)^+$ is energy-dependent. In this case, however, the reaction is driven either by solar radiation (green and blue-green bacteria) or by the proton/charge gradient generated via cyclic electron transfer (purple bacteria). In the blue-green bacteria the extremely high redox potential of the oxygen/water couple necessitates the involvement in non-cyclic electron transfer of a high redox potential photo-pigment system (photosystem II) in addition to the ubiquitous low redox potential system (photosystem I). Since the redox spans of the two photosystems overlap, the excess energy can be used to synthesize a small amount of ATP. Photosystem II contains similar photopigments to those present in photosystem I, plus phycobiliproteins and manganese proteins which assist with the release of oxygen. The oxidation of inorganic reductants in other phototrophs is catalysed by specially adapted cytochromes or novel flavo-cytochromes.

The respiratory chains and photosynthetic electron transfer systems of bacteria, like those of higher organisms, contain both organic redox centres (flavoproteins and quinones) and metal-containing redox centres (cytochromes, iron-sulphur

Bacterial Respiration and Photosynthesis

proteins, molybdoproteins, copper proteins and manganese proteins). The organic centres are of relatively fixed redox potential and transfer hydrogen atoms, whereas the metal centres can often span a wide range of redox potential and catalyse electron and oxygen atom transfer. These two types of redox centre are therefore associated with the oxidation-reduction of organic and inorganic substrates respectively.

A few species of red halobacteria catalyse photophosphorylation in the absence of either bacteriochlorophyll or conventional redox carriers. Instead they use a novel photopigment, bacteriorhodopsin, which is similar in structure to the pigment rhodopsin (visual purple) that functions in vertebrate vision. Bacteriorhodopsin harnesses solar energy directly to the formation of a proton/charge gradient.

Although most photosynthetic bacteria are obligate phototrophs, the halobacteria and some species of purple and blue-green bacteria are facultative and hence catalyse respiratory chain phosphorylation under aerobic conditions. A few species can also grow anaerobically in the dark, conserving energy via substrate-level phosphorylation during the fermentation of glucose or pyruvate.

Respiratory chain and photosynthetic phosphorylation

The free energy released by the redox reactions of respiration and photosynthetic electron transfer is conserved as ATP via the membrane-bound ATP phosphohydrolase. The latter consists of two multipeptide assemblies, termed BF_0 and BF_1 (where BF stands for bacterial coupling factor). BF_1 is a hydrophilic polypeptide complex which is located on the cytoplasmic surface of the energy-coupling membrane and is responsible for ATP synthesis and hydrolysis by the overall complex. It is easily detached from BF_0, but in its soluble form it catalyses only ATP hydrolysis. In contrast, BF_0 is an assembly of hydrophobic polypeptides and proteolipids which forms an intrinsic part of the membrane and probably facilitates the utilization by BF_1 of the proton/charge gradient generated by the various redox systems.

The stoichiometry of ATP synthesis during oxidative and photosynthetic phosphorylation is expressed as the $ATP/O(ATP/2e^-)$ quotient or $P/O(P/2e^-)$ quotient (mole ATP synthesized or phosphate esterified).(g-atom 0 or mole of alternative two-electron acceptor reduced)$^{-1}$.

The determination of the mechanism of membrane-associated energy transduction is one of the major problems of contemporary biochemistry. Although it is now generally accepted that energized protons are the primary intermediates in this process, there is still considerable controversy regarding the generation, location and utilization of these protons. Two major hypotheses, the chemiosmotic and localized proton hypotheses, have been proposed and are currently receiving considerable experimental attention.

The chemiosmotic hypothesis The major tenets of this hypothesis, which was first proposed by Mitchell in 1961, are that oxidative and photosynthetic phosphorylation require (i) a proton-translocating redox system, (ii) a proton-translocating ATP phosphohydrolase, and (iii) a passive coupling membrane which is impermeable to ions, particularly H^+ and OH^-, except via specific exchange-diffusion systems. Energy transduction thus occurs via a proton current which circulates through the insulating membrane and the adjacent bulk aqueous phases (Fig. 1.3a);

Introduction

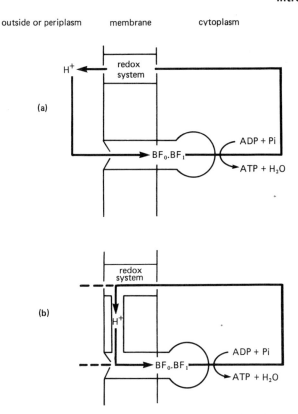

Fig. 1.3 The proton current according to (a) the chemiosmotic hypothesis, and (b) the localized proton hypothesis.

since the latter are in equilibrium, energy storage is transmembrane rather than intramembrane, and takes the form of a delocalized electrochemical potential difference of protons or protonmotive force (Δp or $\Delta_{\bar{\mu}H^+}$; mV). This is variably composed of a chemical potential difference $\Delta pH (pH_{out} - pH_{in})$ and an electrical potential difference or membrane potential ($\Delta \psi$) according to the relationship:

$$\Delta p = \Delta \psi - Z.\Delta pH$$

where $Z (\equiv 2.303\, RT/F)$ has a value of approximately 60 at 25° and serves to convert ΔpH into electrical units. The hypothesis is said to be chemiosmotic since it involves both the transfer of chemical groups (H, H$^-$, e$^-$, O^{2-}) within the membrane and the transport of a solute (H$^+$) across the membrane.

According to the chemiosmotic hypothesis, respiration and photosynthetic electron transfer are described by the general equation:

$$DH_2 + A + zH^+_{(in)} \rightleftharpoons zH^+_{(out)} + AH_2 + D$$

where z is numerically equal to the $\rightarrow H^+O(\rightarrow H^+/2e^-)$ quotient (g-ion H$^+$.

Bacterial Respiration and Photosynthesis

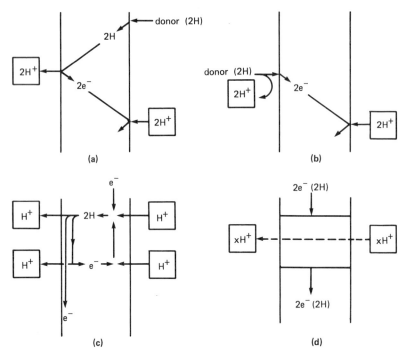

Fig. 1.4 Protonmotive redox reactions. (a) loop, (b) arm, (c) cycle and (d) pump.

g-atom O or mole of alternative two-electron acceptor reduced^{-1}). Redox-linked electrogenic (charge producing) proton translocation is thought to occur at organic centre/metal centre junctions, the centres being spatially and sequentially organized within the membrane such as to catalyse the release of protons at the outer surface and the uptake of protons at the inner surface (Fig. 1.4). These various protonmotive redox arms, loops and cycles may also be replaced by or supplemented with proton pumps in which redox-linked (or, in the case of bacteriorhodopsin, redox-independent) conformational changes in specific membrane proteins lead to changes in the pKa of appropriately located carboxyl or amino groups, and hence to the assymetric uptake and release of protons across the membrane (so-called membrane Bohr effects).

Similarly, ATP synthesis may be described by the equation:

$$H^+ + ADP^{3-} + HPO_4^{2-} + xH^+(\text{out}) \rightleftharpoons xH^+(\text{in}) + ATP^{4-} + H_2O$$

where x is numerically equal to the $\rightarrow H^+/ATP$ ($\rightarrow H^+/P$) quotient (g-ion H^+.mole ATP synthesized or phosphate esterified^{-1}); the other proton is scalar rather than vectorial and reflects ionization changes during the reaction. The ATP/O (ATP/$2e^-$) quotient is therefore equal to the $\rightarrow H^+/O$ ($\rightarrow H^+/2e^-$) quotient divided by the $\rightarrow H^+/ATP$ quotient. Chemiosmosis envisages that ATP synthesis is initiated by the movement of ADP^{3-} and phosphate to the active site of BF_1, with the concomitant release of $2H^+$ into the internal compartment. O^{2-} is then transferred from phosphate to the $2H^+$ which enters from the external compartment via BF_0 under the driving force of the Δp, and the resultant highly active phosphorylium group ($P^+O_3^{2-}$) is attacked by $MgADP^-$ to form ATP^{4-}; the latter, plus water, is finally released into the internal compartment.

The chemiosmotic hypothesis makes several other important predictions with respect to the mechanism of energy transduction: (i) since ATP synthesis (and some other energy-dependent membrane processes) are envisaged to be independent of covalent intermediates and to be driven by a delocalized proton/charge gradient, it should be possible experimentally to drive these reactions at the expense of artificially-imposed ΔpH and/or $\Delta \psi$; (ii) energy transduction should only occur in membranes which effectively separate the external and internal bulk aqueous phases, i.e. the membranes should be topologically closed; (iii) the rate of respiration and photosynthetic electron transfer should be controlled by the back pressure of Δp, thus preventing wasteful and unnecessary redox activity (respiratory control and photosynthetic control); (iv) since Δp is inversely proportional to the rate of proton conductance through the coupling membrane, any compound which increases the latter should thus dissipate Δp and effectively inhibit energy transduction whilst stimulating the redox reactions (such protonophores thus act as uncoupling agents); and (v) in order to prevent respiration and photosynthetic electron transfer from building up an osmotically-disruptive ΔpH, the exchange-diffusion systems in the membrane should catalyse the uptake of protons either with anions or in exchange for cations (i.e. H^+.anion symport and H^+.cation antiport respectively). Such systems, in association with cation uniports, would also facilitate the energy-dependent import and export of metabolically important solutes.

The localized proton hypothesis This hypothesis, which was developed by Williams, differs from chemiosmosis in that it considers the energized protons to be localized (i.e. intramembrane or trans-interface) rather than delocalized (Fig. 1.3b). Furthermore, the postulated absence of a significant osmotic component (i.e. the transport of H^+ across the membrane) means that this hypothesis is satisfied by the presence of a membrane plus a cytoplasmic aqueous phase, whereas chemiosmosis additionally requires the external aqueous phase. The localized proton hypothesis envisages that during respiration and photosynthetic electron transfer, protons and electrons are separated at organic centre/metal centre junctions, and that subsequent proton diffusion between the redox and ATP phosphohydrolase systems is not only extremely rapid but also under strict kinetic control, such that the protons equilibrate only relatively slowly with the adjacent aqueous phases. Redox-linked conformational changes in $BF_0.BF_1$ are thought to exclude water from the active site of the ATP phosphohydrolase and allow the controlled access of phosphate, ADP and protons, the binding of the protons leading to further conformational changes which facilitate the removal of water from ADP and phosphate, and hence the synthesis of ATP. This hypothesis thus stresses the importance of water in oxidative and photosynthetic phosphorylation, and emphasizes the need for proton-binding rather than proton transport (the absence of osmosis implies that ATP synthesis should be possible in non-vesicular membrane preparations). Furthermore, it makes no predictions with respect to the stoichiometry of proton movement (i.e. $\rightarrow H^+/2e^-$ and $\rightarrow H^+/ATP$ quotients) and claims that the only obligatory stoichiometry is that which is observed experimentally for overall ATP synthesis (i.e. the $ATP/2e^-$ quotient). Reversed electron transfer and uncoupling of

Bacterial Respiration and Photosynthesis

energy transduction are thought to occur via somewhat analogous mechanisms to those of chemiosmosis, the former being driven by the back pressure of the localized proton concentration, the intramembrane dissipation of which by protonophores leads to uncoupling.

Since the results of most studies of bacterial respiration and photosynthesis have been interpreted almost entirely in terms of chemiosmosis, the latter will be used in this book as the working mechanism of membrane energy transduction. It will be seen, however, that although some systems are probably chemiosmotic, others can be interpreted just as well in terms of the localized proton hypothesis, and in a few cases the latter appears to explain the experimental observations more successfully.

The coupling membrane

The ability of most bacteria to react positively or negatively to the Gram stain reflects basic differences in the structure and composition of their cell envelope. In Gram positive bacteria, all of which are chemoheterotrophs, the cytoplasm is usually bounded by an 8 μm thick membrane (the plasma membrane or cytoplasmic membrane) that is responsible both for energy coupling and for the compartmentation and transport of solutes. This membrane is surrounded by, and tightly connected to, a 10 to 80 μm thick rigid cell wall which is composed principally of peptidoglycan and is occasionally covered with a layer of slime.

The cell envelope of Gram negative bacteria (i.e. many chemoheterotrophs, especially those with specialized metabolic properties, and all chemolithotrophs and phototrophs) is a very much more complex structure which consists of three layers: an outer membrane, a thin (2 μm) and net-like peptidoglycan layer, and an inner membrane. The first two form a relatively rigid exoskeleton which is separated from the more flexible inner membrane by the periplasmic space. The 8 μm thick outer membrane, which is composed of phospholipid abundantly interspersed with protein, lipoprotein and lipopolysaccharide, has a limited capacity to control solute uptake; it thus provides a barrier to potentially toxic materials such as detergents and antibiotics, whilst allowing the free passage of many small molecules either by simple diffusion or via transmembrane pores. The periplasmic space is 3 to 4 μm wide and contains various hydrolytic enzymes, solute-binding proteins and, less frequently, certain redox carriers. The inner (plasma) membrane has similar structural and functional properties to its counterpart in Gram positive bacteria.

In many organisms the plasma membrane simply follows the inner contours of the cell, but in others it exhibits a more complex shape and can extend into the cytoplasm to form the intracytoplasmic membrane. Extensive vesicular and lamellar intrusions of various types, including layers of concentric or parallel membranes, are exhibited by some obligately aerobic chemoheterotrophs (particularly those which grow on one-carbon compounds or which fix atmospheric nitrogen) and by many chemolithotrophs and purple phototrophs. Relatively large closed vesicles (chlorosomes; originally called chlorobium vesicles) are found in green phototrophs, where they are arranged around the periphery of the cytoplasm and may be loosely attached to the plasma membrane. Uniquely, a second and quite separate intracytoplasmic membrane is dispersed throughout the cytoplasm of the blue-green bacteria where it is organized into stacks of flattened sacs (thylakoids) that are covered by small granules (phycobilisomes) which contain the light-harvesting apparatus.

Introduction

In most bacteria the coupling membrane consists of a classical tail-to-tail phospholipid bilayer interspersed with protein. The phospholipids are mostly phosphatidylglycerol and phosphatidylcholine in Gram-positive bacteria, and phosphatidylethanolamine and phosphatidylcholine in Gram-negative organisms. The coupling membrane functions optimally in a fairly fluid liquid crystalline state, and its composition is therefore geared to maintaining this state. Thus, since the fluidity of the membrane is chiefly determined by the melting points of the fatty acid components of the phospholipids, the membranes of bacteria which grow at high temperatures (thermophiles) contain mainly longer chain saturated fatty acids with high melting points, whereas the converse is true in low temperature organisms (psychrophiles). Furthermore, bacteria which grow at extremes of salinity, pH and/or temperature (i.e. halophiles, acidophiles, aklaliphiles, thermoacidophiles and thermoalkaliphiles) contain significant amounts of unusual phospholipids

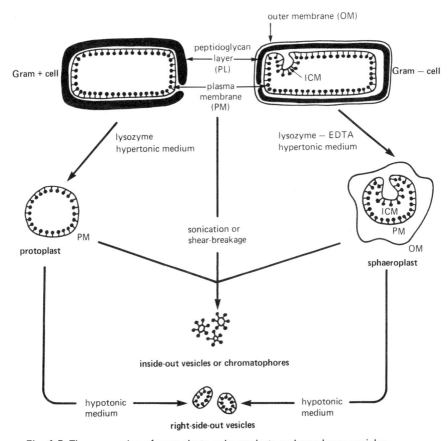

Fig. 1.5 The preparation of protoplasts, sphaeroplasts and membrane vesicles. Abbreviations: OM, outer membrane; PL, peptidoglycan layer; PM, plasma (inner) membrane; ICM, intracytoplasmic membrane (shown here as a single vesicular intrusion into a phototrophic cell). After Konings (1977).

and/or more complex lipids (glycolipids, sulpholipids, diglycerol tetra-ethers). Indeed, some thermoacidophiles go so far as to abandon the bilayer membrane in favour of a complex-lipid monolayer. In each case the alteration in membrane composition and/or structure reflects an attempt to maintain the required fluidity and integrity in the face of a hostile environment. The protein components of the coupling membrane are either predominantly hydrophilic proteins which are associated with the periphery of the membrane (extrinsic proteins, e.g. BF_1 and some redox carriers) or predominantly hydrophobic proteins which are integral components of the membrane and sometimes completely span the latter (intrinsic proteins, e.g. BF_0, transport permeases and most redox carriers).

Exposure of whole cells to the enzyme lysozyme under the appropriate conditions causes degradation of the peptidoglycan layer and hence the conversion of Gram-positive and Gram-negative bacteria into osmotically fragile protoplasts and sphaeroplasts respectively (Fig. 1.5). Both of these have been used to investigate energy transduction, but more recently greater emphasis has been placed on the use of membrane vesicles. The latter are usually prepared either by exposing protoplasts and sphaeroplasts to ultrasound or hypotonic conditions, or by shearing whole cells under pressure (e.g. using a French pressure cell). The vesicles produced by osmotic shock are relatively large (0.8 to 1.1 μm diameter), do not readily hydrolyse ATP, and are called right-side-out vesicles since the orientation of the membrane is the same as in the parent protoplasts or sphaeroplasts, i.e. BF_1 is on the inner surface. In contrast, the vesicles prepared by sonication or shearing are smaller (≤ 0.1 μm diameter), readily hydrolyse ATP, and are termed inside-out vesicles (or chromatophores when prepared from phototrophs) since the membrane is oriented in the opposite direction to that in the intact cell, i.e. BF_1 is exposed on the outer surface. Inside-out vesicles are routinely used for investigating oxidative and photosynthetic phosphorylation, whereas right-side-out vesicles are more suited to the study of solute transport. It should be noted, however, that such vesicles are often far from being structurally perfect or homogeneous, particularly when prepared from respiratory membranes.

Summary

Bacteria conserve energy mostly in the form of adenosine 5'-triphosphate (ATP) which they synthesize via substrate-level, oxidative (respiratory chain) or photosynthetic phosphorylation. The former is principally associated with anaerobic fermentative bacteria and is a scalar cytoplasmic process that entails the sequential formation of non-phosphorylated and phosphorylated, energy-rich intermediates. The latter two processes, on the other hand, are the major energy-conserving reactions in non-fermentative organisms, and are vectorial membrane-bound phenomena in which sequential redox reactions drive ATP synthesis via the intermediate formation and utilization of energized protons.

During respiration and photosynthesis reducing equivalents are transferred from diverse donors to higher redox potential acceptors via a highly organized sequence of redox carriers (nicotinamide nucleotides, flavoproteins, iron-sulphur proteins, quinones, cytochromes, copper proteins and molybdoproteins) which contain organic or metal redox centres. Oxygen is the most frequently used acceptor for respiration, but is occasionally replaced by alternative inorganic or organic oxidants, and organic donors are sometimes replaced by inorganic reductants.

Introduction

Energy conservation during photosynthesis arises principally from cyclic electron transfer between endogenous reductants and oxidants (chlorophylls and cytochromes respectively) that are generated by the initial photoreaction. Non-cyclic electron transfer occurs from exogenous organic or inorganic reductants to $NAD(P)^+$, thus generating $NAD(P)H$ for carbon dioxide assimilation. In a few highly specialized phototrophs, energy conservation is entirely independent of electron transfer.

Two major hypotheses seek to explain the mechanism of oxidative and photosynthetic phosphorylation. The chemiosmotic hypothesis envisages that a proton current circulates through the insulating membrane and the adjacent bulk aqueous phases, and that energy storage is effected by a delocalized protonmotive force (Δp). In contrast, the localized proton hypothesis proposes that the protons are confined within the membrane, or at the membrane-cytoplasm interface, and thus stresses proton binding rather than proton transport. The two hypotheses envisage significant differences in the way in which the energized protons drive ATP synthesis via the ATP phosphohydrolase.

Energy coupling occurs in the cytoplasmic or intracytoplasmic membrane, which is highly impermeable to H^+, OH^- and other ions except via specific porter systems. During growth at extremes of temperature, pH or salinity, the fatty acid and phospholipid composition of the coupling membrane undergoes extensive changes in an attempt to maintain this property.

References

GARLAND, P. B. (1977). Energy transduction in microbial systems. In: *Microbial Energetics* pp. 1–21. Edited by B. A. Haddock and W. A. Hamilton. Society for General Microbiology Symposium 27. Cambridge University Press, Cambridge.

HAROLD, F. M. (1978). Vectorial metabolism. In: *The Bacteria* Vol. 6 pp. 463–521. Edited by L. N. Ornston and J. R. Sokatch. Academic Press, New York.

JONES, C. W. (1981). *Biological Energy Conservation: Oxidative Phosphorylation* (2nd Edition). Outline Studies in Biology series. Chapman and Hall, London.

KONINGS, W. N. (1977). Active transport of solutes in bacterial membrane vesicles. *Advances in Microbial Physiology* 15: 175–253.

LEHNINGER, A. L. (1975). *Biochemistry*. (2nd Edition). Worth Publishers, New York.

MITCHELL, P. (1979). David Keilin's respiratory chain concept and its chemiosmotic consequences. In: *Les Prix Nobel en 1978* pp. 135–72. Nobel Foundation, Stockholm.

SALTON, M. J. R. and OWEN, P. (1976). Bacterial membrane structure. *Annual Review of Microbiology* 30: 451–82.

STANIER, R. Y., ADELBURG, E. A. and INGRAHAM, J. L. (1977). *General Microbiology* (4th Edition). Prentice Hall, New Jersey.

THAUER, R. K., JUNGERMANN, K. and DECKER, K. (1977). Energy conservation in chemotrophic anaerobic bacteria. *Bacteriological Reviews* 41: 100–80.

WILLIAMS, R. J. P. (1978). The multifarious couplings of energy transduction. *Biochimica et biophysica Acta* 505: 1–44.

2 Aerobic respiration in chemoheterotrophs and facultative phototrophs

A taxonomically wide range of chemoheterotrophic and facultatively phototrophic bacteria respire aerobically. According to species, they oxidize various organic reductants which arise either as intermediates of intracellular metabolism or are provided by the growth environment (e.g. NADPH, NADH, succinate, lactate, malate and methanol), with the concomitant reduction of molecular oxygen to water or, very occasionally, to hydrogen peroxide.

Respiratory chain composition

The aerobic respiratory chains of bacteria contain the same basic types of redox carriers as those present in the mitochondria of eukaryotes, i.e. flavoproteins (Fp) and iron-sulphur proteins (Fe-S) (which comprise the dehydrogenases), quinones, cytochromes and cytochrome oxidases.

The iron-sulphur flavoprotein dehydrogenases catalyse the oxidation of NADH, succinate and glycerol-3-phosphate, and occasionally also NADPH, formate, lactate and malate (although in many species of bacteria, particularly facultative anaerobes, the latter two substrates are often additionally oxidized by soluble, NAD^+-linked dehydrogenases).

Flavoproteins These consist of an apoprotein of varied molecular weight plus a tightly-bound prosthetic group, either flavin mononucleotide (FMN) or flavin adenine dinucleotide (FAD). The flavins carry a maximum of two hydrogen atoms in their three-ringed isoalloxazine nucleus, and on reduction lose their distinct yellow colour ($\lambda_{max} \simeq 450$ nm) and become virtually colourless. The E'_0 values of the free $FMN/FMNH_2$ and $FAD/FADH_2$ couples are approximately -205 and -219 mV respectively, but these values are often substantially modified when the flavin is bound to the protein. FAD is normally bound covalently via a histidyl residue (8-N3 histidyl-FAD) and hence operates at a relatively high redox potential (≥ 0 mV), whereas FMN is usually bound ionically via its negatively charged phosphate groups and operates at a much lower redox potential (≤ -100 mV).

Iron-sulphur proteins These are relatively small (MW $\leq 30,000$) proteins which contain either 2,4 or 8 atoms of iron plus the same amount of labile-sulphur. The [2Fe-2S] proteins are characterized by a relatively simple redox centre in which the two iron atoms are held in a lattice of four atoms of cysteine-sulphur plus two atoms of labile-sulphur (so-called because it is released as hydrogen sulphide under acid conditions). In contrast, the [4Fe-4S] proteins contain a more complex redox centre in which the four iron atoms from a cube with four atoms of labile-sulphur, the cube being held in place by four cysteines; the [8Fe-8S] proteins consist of two separate [4Fe-4S] centres. On reduction, the [2Fe-2S] and [4Fe-4S] proteins accept one electron, and the 2[4Fe-4S] proteins accept two, one at each redox centre. Since the iron-sulphur proteins absorb only rather weakly at 450 to 500 nm in their oxidized

Aerobic respiration in chemoheterotrophs and facultative phototrophs

forms, and even less so when reduced, their redox changes are usually monitored using electron paramagnetic resonance (EPR) spectroscopy. Using this technique, their E_m values (the nearest equivalent to E'_0 which can be measured for membrane-bound redox carriers) range from -600 to $+350$ mV, but the precise reasons for this wide variation have yet to be resolved. The 2[4Fe-4S] proteins are restricted to fermentation reactions, and the very low redox potential iron-sulphur proteins ($E_m < -250$ mV) are predominantly associated with photosynthetic electron transfer rather than respiration.

Physical analyses of membrane-bound respiratory chains have indicated the presence of FMN plus four low redox potential Fe-S centres in the NADH dehydrogenases of *Paraccocus denitrificans* and *Escherichia coli*, and FAD plus three Fe-S centres have been detected in the succinate dehydrogenase of the latter organism. In each case the Fe-S centres closely resemble those present in the same enzymes in eukaryote mitochondria (centres N-1 to N-4, and S-1 to S-3 respectively) and include both [2Fe-2S] and [4Fe-4S] proteins.

Most of the respiratory chain dehydrogenases are tightly embedded in the coupling membrane and hence are difficult to extract, particularly in forms which retain their *in vivo* properties. However, several relatively simple flavoprotein dehydrogenases have recently been purified from the aerobic respiratory chain of *E. coli*, including D-lactate dehydrogenase (MW 75000; FAD) and L-lactate dehydrogenase (MW 480000; FMN). Succinate dehydrogenase has been purified from *Rhodospirillum rubrum* (MW 97000; FAD and Fe-S). In contrast, methanol dehydrogenase is only loosely bound to the respiratory membranes of methylotrophs such as *Methylophilus methylotrophus* and *P. denitrificans* and can be purified fairly easily (MW 60000 to 160000). It contains neither flavin nor Fe-S centres, and its redox component appears to be a quinone (interestingly, a similar prosthetic group is present in the soluble glucose and methylamine dehydrogenases of various organisms).

Quinones Most respiratory chain dehydrogenases transfer reducing equivalents to either ubiquinone (Q) or menaquinone (MK; vitamin K_2). The former is generally the acceptor in Gram negative bacteria, and the latter in Gram positives; a few facultative anaerobes such as *E. coli* contain both quinones, but tend to use ubiquinone for aerobic respiration. Each quinone consists of a substituted 1,4-benzoquinone (Q) or 1,4,-naphthoquinone (MK) nucleus attached to a long polyisoprenoid side chain. A few species of *Streptococcus* and *Haemophilus* contain demethylmenaquinones (DMK) which lack the methyl group on the nucleus, and many organisms contain small amounts of partially hydrogenated quinones in which one or more of the double bonds in the side chain is reduced. All of these quinones are lipids, and on reduction accept 2H to form the corresponding quinol, possibly via a semiquinone intermediate. Unlike the other respiratory chain components, they absorb in the near ultraviolet region of the spectrum and their redox behaviour *in situ* is therefore only monitored with difficulty. However, such experiments have yielded E_m values of $+113$, $+36$ and -74 mV respectively for Q, DMK and MK, and have indicated that the quinols transfer electrons to the terminal cytochrome system. The central positions of these redox carriers in the respiratory chain have been confirmed by extraction-reactivation experiments in which the quinones are removed from lyophilized membranes using non-polar organic solvents, with the concomitant loss of respiratory activity, and then added back to the depleted membranes with the resultant reconstitution of this activity.

Bacterial Respiration and Photosynthesis

Table 2.1 Cytochromes of aerobic respiratory chains

Cytochrome	Prosthetic group	Inhibition by: CO	Inhibition by: CN	E_m (mV)	Oxidase capacity	Purification	MW
b	haem b	−	−	−104 to +110	−	±	12 000 – 17 500
c	haem c	−	−	+190 to +342	−	+	12 000 – 100 000
aa_3	2 haem a	++	++	+200 to +265 and +360 to +375	+	+	73 000
o	2 Cu, 2 haem b	++	+	+210, −122 to +417	+	+	28 000
d	2 haem d	++	±	+280	+	±	350 000
a_1	2 haem b, 2 haem a	±, ++	++, ++	+140 and +250; +160 and +260	−/+, −(?)	−, +	12 500
c_{co}	haem c	++	++	+360			
ca	1 haem c, 1 to 2 haem a, 1 Cu	++	++	?	?	±	38 000 – 118 000

Aerobic respiration in chemoheterotrophs and facultative phototrophs

The unusual hydrophilic quinone that comprises the prosthetic group of methanol, glucose and methylamine dehydrogenases (4,5-pyrroloquinoline quinone or methoxatin, $E_m PQQ/PQQH_2 \simeq +150\,mV$) also transfers reducing equivalents directly to the terminal cytochrome system.

Cytochromes Four types of cytochrome are known (a, b, c and d), each of which consists of one or more haem prosthetic groups bound to an apoprotein. The haem is composed of porphyrin (four pyrrole rings joined by $=CH-$ bridges) plus a central iron atom which on reduction accepts a single electron, i.e. $Fe^{3+} + e^- \rightleftharpoons Fe^{2+}$. It is likely, however, that some b-type cytochromes are also able to carry a proton. The four types of cytochrome differ principally in the nature of their substituent groups at the periphery of the haem. The iron generally forms an octahedral coordination complex with six ligands; four of these are the nitrogen atoms of the pyrroles, which serve to hold the iron roughly within the plane of the porphyrin ring, and the other two are suitable atoms in adjacent amino acid residues (e.g. histidyl-N, methionyl-S). However, in those cytochromes which are rapidly autoxidizable, and hence can act as oxidases, one of these latter two axial positions can be occupied by oxygen, water or an inhibitor such as carbon monoxide. The individual cytochromes exhibit a very wide range of chemical and physical properties (Table 2.1). Thus their E_m values and long wavelength absorption maxima, both of which are dictated by the structures and binding properties of the haems, range from approximately -100 to $+400\,mV$ and from approximately 550 to 650 nm respectively, and their molecular weights are between the 12000 of some b- and c-type cytochromes and the 350000 of cytochrome oxidase d. Many c-type cytochromes, and at least one b-type cytochrome, are extremely hydrophilic and hence are often only loosely attached to the membrane or are even present in the periplasm, whereas most of the b-type cytochromes and oxidases are hydrophobic intrinsic membrane proteins. $E.\ coli\ b_{562}$ and c-type cytochromes from several different organisms have recently been purified to homogeneity and their complete structures determined by X-ray diffraction analyses.

The cytochrome systems of aerobic chemoheterotrophs and facultative phototrophs usually consist of several non-autoxidizable b- and c-type cytochromes, plus one or more cytochromes which bind carbon monoxide and hence may be regarded as potential terminal oxidases, i.e. cytochromes aa_3, o, d, a_1, c_{co} and possibly ca (although the latter, which has been extracted from several thermophiles, may simply be an impure or specialized form of aa_3). Kinetic analyses using rapidly-resolving highly sensitive spectroscopic techniques have so far confirmed an oxidase function only for cytochromes aa_3, o and d, although cytochrome a_1 appears to be kinetically competent to act as an oxidase in some chemolithotrophs. There is some evidence that these oxidases, like cytochrome oxidase aa_3 from mitochondrial respiratory chains, contain up to four one-electron redox centres (i.e. a combination of similar or different haems, or of haem plus copper) which enable them to catalyse the four-electron reduction of molecular oxygen to water. Although relatively few cytochrome oxidases have been extensively purified, there is evidence that most of them are oligomeric proteins (i.e. they contain several subunits) and that some of them are large enough to act as transmembrane electron carriers.

Several genera of aerobic and facultatively anaerobic chemoheterotrophs, which include the *Streptococcaceae* and *Lactobacillaceae*, do not normally synthesize cytochromes. When grown aerobically, their respiratory chains are terminated by a

Bacterial Respiration and Photosynthesis

flavin oxidase which reduces molecular oxygen to hydrogen peroxide. The potentially toxic effects of the latter are obviated by the action of catalase:

$$2H_2O_2 \longrightarrow 2H_2O + O_2$$

or peroxidase:

$$XH_2 + H_2O_2 \longrightarrow 2H_2O + X$$

At least one species of *Streptococcus* synthesizes a complete cytochrome-containing respiratory chain when its growth medium is supplemented with haematin.

Transhydrogenase Many chemoheterotrophs and facultative phototrophs contain a membrane-bound nicotinamide nucleotide transhydrogenase which catalyses the reversible transfer of a hydride ion between NADPH (NADP$^+$) and NAD$^+$ (NADH) according to the equation:

$$NADPH + H^+ + NAD^+ \rightleftharpoons NADH + H^+ + NADP^+$$

The enzyme has a molecular weight of approximately 94000 and does not appear to contain any redox carriers The equilibrium constant of the transhydrogenase reaction would normally be close to unity since the E'_θ values of the two redox couples differ by only 4 mV. However, in most organisms the enzyme is energy-linked (i.e. ATP or the protonmotive force generated by respiration enhances both the rate and extent of NADP$^+$ reduction) and hence directs reducing equivalents from the predominantly NAD$^+$-linked dehydrogenases of catabolism towards the predominantly NADPH-oxidizing enzymes of biosynthesis. Some organisms in the genus *Pseudomonas* contain soluble or membrane-bound transhydrogenases which are energy-independent; the major function of these enzymes, many of which contain flavin, is probably to ensure that excess reducing equivalents generated by NADP$^+$-linked growth substrates are transferred to NAD$^+$ and subsequently oxidized via NADH oxidase.

The aerobic respiratory chains of chemoheterotrophs and facultative phototrophs, like those of eukaryote mitochondria, are potentially sensitive to a wide variety of classical electron transfer inhibitors which bind to, and hence block the action of, specific redox centres. Thus many NADH dehydrogenases are inhibited by rotenone and piericidin A, succinate dehydrogenases by malonate and carboxin, and the central quinone-*b* regions by antimycin A and 2-*n*-heptyl-8-hydroxyquinoline-N-oxide (HQNO); many cytochrome oxidases are often sensitive to carbon monoxide, cyanide and azide. However, a close comparison of these bacterial and mitochondrial electron transfer systems reveals that this apparent unity in redox carrier composition and inhibitor sensitivity is largely superficial and that only a very few species of bacteria contain aerobic respiratory chains which are very similar to those of mitochondria, e.g. *P. denitrificans, Alcaligenes eutrophus* (*Hydrogenomonas eutropha*) and *Rhodopseudomonas sphaeroides*. Indeed, it has been proposed (partly on the basis of similarities in redox carrier composition, sensitivities to electron transfer inhibitors, and other membrane properties) that the inner membrane of the present-day mitochondrion may have evolved from the plasma membrane of an ancestral relative of *P. denitrificans* via endosymbiosis of the latter with a primitive host cell.

Species differences Large variations in redox carrier composition and sensitivity to electron transfer inhibitors occur between different species of chemoheterot-

Aerobic respiration in chemoheterotrophs and facultative phototrophs

rophic and facultatively phototrophic bacteria; indeed, one of the major characteristics of these aerobic respiratory chains is their immense variety (Table 2.2). These variations fall into three major categories: (i) the replacement of one redox carrier by another with basically similar properties, e.g. the replacement of ubiquinone by menaquinone (both are lipophilic hydrogen carriers) or the replacement of one type of cytochrome oxidase by another (all are autoxidizable electron carriers); (ii) the replacement of one redox carrier by another with significantly different properties, e.g. the replacement of an energy-dependent transhydrogenase by an energy-independent transhydrogenase; and (iii) the addition or deletion of a limited number of redox carriers, e.g. transhydrogenases of either type and/or cytochrome c.

These variations generally have little effect on the respiratory capacities of isolated membrane preparations, although some of them can significantly influence the efficiency of oxidative phosphorylation or may be accompanied by altered sensitivities to classical respiratory chain inhibitors. Thus, many respiratory systems (particularly those present in the *Pseudomonadaceae, Enterobacteriaceae* and closely related families) are resistant to piericidin A, rotenone and antimycin A, and those systems which contain cytochrome oxidase d are relatively insensitive to cyanide and azide.

Phenotypic modification of respiratory chains Qualitative and quantitative variations in respiratory chain composition can also occur within a single species of bacterium. These usually reflect changes in the growth environment, particularly (i) the decreased availability of essential nutrients, (ii) the presence of respiratory chain inhibitors, or (iii) alterations in the nature of the carbon/energy source. Thus, growth in batch or continuous culture under oxygen-limited conditions (Table 2.2) is often accompanied by the increased synthesis of alternative cytochrome oxidases, e.g. d (plus a_1) relative to either o or aa_3, or o relative to aa_3. Such changes undoubtedly reflect attempts by these organisms to combat an insufficiency of oxygen through the differential synthesis of alternative oxidases which have higher affinities for this electron acceptor. This conclusion is generally supported by the results of experiments designed to determine the relative or absolute affinities of different oxidases for molecular oxygen, i.e. via growth competitions under oxygen-limited non-fermentative conditions or via direct estimations of K_m values respectively (affinities are in the order $d > aa_3$ or o, the K_m for oxygen being $\leq 1\,\mu M$).

Respiratory chain phenotypes can also be altered by growing organisms in continuous culture such that a single nutrient which is essential for the biosynthesis of a particular redox carrier becomes rate-limiting for growth. In the few species of bacteria which have so far been examined in this manner, which include *E. coli* and *P denitrificans*, iron-limitation leads to a fall in the concentrations of membrane-bound iron-sulphur proteins and cytochromes. Sulphate-limitation also decreases the content of quinones and iron-sulphur proteins (although the extent of this loss varies considerably between species) but appears to stimulate the synthesis of alternative cytochrome components; the reason for its effect on quinone and cytochrome concentrations is unclear. Since there is some evidence that bacterial cytochrome oxidase aa_3 is associated with copper, like its counterpart in eukaryote mitochondria, copper-limited continuous culture is a potentially powerful tool for investigating cytochrome oxidase functions. This technique has been applied successfully to yeast but not, as yet, to bacteria.

Table 2.2 Simplified respiratory chain compositions following heterotrophic growth under excess or limiting oxygen concentrations

Oxygen concentration	Organism	Respiratory chain composition						
Excess	*Paracoccus denitrificans*	Th	Ndh	Q		b	c	aa_3	(o)
	Alcaligenes eutrophus	Th	Ndh	Q		b	c	aa_3	(o)
	Rhodopseudomonas sphaeroides	Th	Ndh	Q		b	c	aa_3	(o)
	Arthrobacter globiformis		Ndh	Q		b	c	aa_3	(o)
	Micrococcus lysodeikticus	Th	Ndh		MK	b	c	aa_3	(o)
	Pseudomonas fluorescens	Th	Ndh	Q		b	c		o
	Escherichia coli	(Th)	Ndh	Q	(MK)	b			o
	Klebsiella pneumoniae	(Th)	Ndh	Q		b			o
	Vibrio (Beneckea) natriegens	Th*	Ndh		MK	b	c		o c_{co} $a_1 d$
	Bacillus megaterium		Ndh			b		aa_3	(o)
	Azotobacter vinelandii		Ndh	Q		b	c		o c_{co} $a_1 d$
	Methylophilus methylotrophus		Ndh	Q	PQQ	b	c	aa_3	o
Limiting	*Paracoccus denitrificans*	Th	Ndh	Q		b	c	aa_3	o
	Alcaligenes eutrophus	Th	Ndh	Q		b	c	aa_3	o
	Arthrobacter globiformis		Ndh	Q		b	c	(aa_3)	(o) d
	Escherichia coli	(Th)	Ndh	Q	MK	b			(o) $(a_1) d$
	Klebsiella pneumoniae	(Th)	Ndh	Q		b			(o) $(a_1) d$

Abbreviations: Th, nicotinamide nucleotide transhydrogenase; Ndh, NADH dehydrogenase; Q, ubiquinone; MK, menaquinone; cytochromes are referred to by letter. Note that more than one b- or c-type cytochrome is often present within a single organism. Brackets indicate redox carriers of low concentration or activity; the asterisk denotes an energy-independent transhydrogenase.

Aerobic respiration in chemoheterotrophs and facultative phototrophs

Similar changes in cytochrome oxidase content to those induced by oxygen-limitation can also occur during growth in the presence of low concentrations of cyanide, achieved by adding cyanide to the medium or by culturing organisms such as *Chromobacterium violaceum* under conditions which stimulate cyanogenesis, thus confirming conclusions from *in vitro* experiments that the alternative oxidases have lower sensitivities to this inhibitor (the K_i for cyanide of cytochrome d can be as high as 10 mM compared with ≤ 0.3 mM for o and $\leq 20 \mu$M for aa_3). The exposure of continuous cultures of *P. denitrificans* to rotenone leads to the selection of a rotenone-resistant strain which appears to lack the N-2 iron-sulphur protein in its NADH dehydrogenase. There is some evidence that both this lesion and the one induced by sulphate-limitation lead to the loss of energy coupling in this region of the respiratory chain.

The replacement of glycerol or succinate in the *E. coli* growth medium by a fermentable and more rapidly metabolizable carbon source such as glucose, which is a powerful catabolite repressor, causes a general decrease in the concentrations of existing redox components which can be reversed by the addition of cAMP to the growth medium. In contrast, the presence of high concentrations of amino acids leads to the repression of specific redox carriers such as transhydrogenase and, in some organisms, cytochrome oxidase aa_3, but the physiological basis of this amino acid effect is currently unclear.

Genotypic modification of respiratory chains The redox carrier composition of bacterial respiratory chains can also be manipulated genetically, with the formation of mutant strains which are deficient in one or more redox centres. The most popular and rewarding organisms to use for this type of work are undoubtedly *E. coli* K12 and related strains since (i) they are very amenable to mutagenesis and to sophisticated genetic analysis, (ii) they readily allow the transfer of mutant alleles via phage or plasmid vectors, and (iii) they are facultative anaerobes and thus can obtain energy for growth using either substrate-level or oxidative phosphorylation. This latter property facilitates the selection and screening of putative electron transfer mutants since these grow readily on fermentable carbon sources such as glucose, but not on a mixture of non-fermentable compounds such as succinate + malate + acetate.

Most of the aerobic respiratory chain mutants of *E. coli* which have been selected in this way are defective in their ability to synthesize particular redox centres (e.g. ubiquinone, menaquinone, haem), apoproteins (e.g. of various dehydrogenases) or transport systems for the uptake of essential precursors (e.g. the iron-chelator enterochelin, the absence of which leads to iron deficiency). Functional electron transfer activity can often be restored to these mutants by incubating whole cells or depleted membranes with an easily assimilated form of the missing redox carrier or one of its precursors; for example, aerobic respiration can be restored to membranes from selected Ubi$^-$ and Hem$^-$ mutants of *E. coli* by the addition of a hydrophilic ubiquinone homologue $(Q-1)$ and haematin + ATP respectively, confirming that ubiquinone and *b*-type cytochromes are obligatory components of the aerobic respiratory chain of this organism. The phenotypic and genotypic modification of the aerobic respiratory chain of *E. coli* is summarized in Fig. 2.1.

Cytochrome *c*-deficient mutants have been isolated from several obligately aerobic chemoheterotrophs (*A. vinelandii*, *P. denitrificans* and *Pseudomonas* AM1) and from the facultative phototroph *R. capsulata*. Such mutants are easy to screen

Bacterial Respiration and Photosynthesis

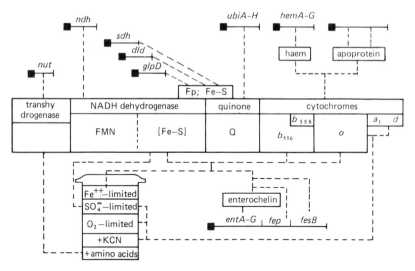

Fig. 2.1 Phenotypic and genotypic modification of the aerobic respiratory chain of
E. coli (courtesy of Dr. B.Å. Haddock). Genes which are responsible for redox carriers
are shown in italics (e.g. *ndh*, NADH dehydrogenase), gene products are enclosed in
square boxes, and the different growth conditions which give rise to phenotypic
changes are listed in the culture vessel.

since, unlike the parent organisms, they exhibit a negative Nadi or Oxidase reaction:

$$\alpha\text{-naphthol} + N, N'\text{-dimethyl-}p\text{-phenylenediamine} + O_2 \xrightarrow[+ \text{cytochrome oxidase}]{\text{cytochrome } c} \text{indophenol blue} + H_2O$$

and hence fail to turn blue. All four organisms retain their ability to oxidize succinate or NAD^+-linked substrates, but the facultative methylotrophs *P. denitrificans* and *P.* AM1 are no longer able to oxidize methanol, and *R. capsulata* will not grow photosynthetically. A different mutant of *R. capsulata* has been isolated by the same screening procedure, which contains normal amounts of cytochrome *c* but lacks one of its terminal oxidases; it exhibits the same properties as the cytochrome *c*-deficient mutant, except that it continues to grow photosynthetically. These results indicate that cytochrome *c* is required for methanol oxidation and cyclic photosynthetic electron transfer, and that all of these organisms contain a branched respiratory chain, one branch of which is cytochrome *c*-independent.

Pathways of respiration

Chemoheterotrophic and facultatively phototrophic bacteria exhibit considerable variations in the complexity and organization of their aerobic respiratory

Aerobic respiration in chemoheterotrophs and facultative phototrophs

pathways. All of them exhibit extensive branching at the level of their primary dehydrogenases, thus allowing reducing equivalents from various substrates of widely differing E'_θ values and coenzyme/prosthetic group specificities to be channelled into a common respiratory chain. In contrast, the terminal cytochrome system may be either branched or linear (Fig. 2.2). Linear terminal pathways are relatively rare and are limited to the few systems which contain only one functional cytochrome oxidase (e.g. *E. coli* cultured under highly aerobic conditions).

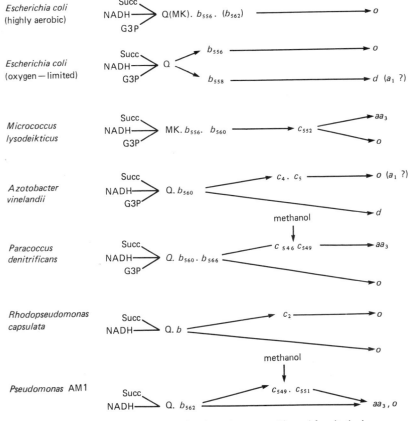

Fig. 2.2 Aerobic respiratory pathways in chemoheterotrophic and facultatively phototrophic bacteria. Cytochromes are referred to by letter plus, where possible with *b*- and *c*-type cytochromes, their wavelength maxima in low temperature reduced minus oxidized difference spectra; c_2 is the trivial abbreviation of the cytochrome which is present in many species of *Rhodospirillaceae* (chapter 4). Brackets indicate redox carriers of low concentration or activity. Note that (i) cytochrome *o* becomes the major oxidase in cytochrome *c*-deficient mutants of *Pc. denitrificans*, and (ii) *Rps. capsulata* contains two active cytochromes *o* with quite different E_m values.

Bacterial Respiration and Photosynthesis

Branching is usually associated with the presence of more than one functional species of cytochrome oxidase and, in its simples form, consists of electron transfer from penultimate non-autoxidizable b- or c-type cytochromes to molecular oxygen via two or more oxidases (e.g. oxygen-limited *E. coli*, and probably also *Micrococcus lysodeikticus*). In a few respiratory chains branching is more complex and the separate terminal pathways appear to contain b- and/or c-type cytochromes, and often different cytochrome oxidases as well (e.g. *A. vinelandii*, *V. natriegens*, *P. denitrificans*, *R. capsulata* and *P.* AM1, and also a wide range of organisms cultured under oxygen limited conditions). There is increasing evidence that in these complex systems one branch contains cytochrome c and is terminated by cytochrome oxidases aa_3 or o, whereas in the other branch a b-type cytochrome donates electrons directly to cytochrome oxidase d, aa_3 or o.

The control of respiration

Bacterial respiration is subject to both fine and coarse control. The former is exerted by the inherent kinetic properties of the respiratory chain components and by the backpressure of the energized protons, and the latter by the repression/induction of redox carrier biosynthesis as already discussed.

The rate-limiting step of the respiratory chain appears almost invariably to be at the level of the primary dehydrogenases rather than in the quinone-cytochrome system, and thus reflects the physiological necessity for several dehydrogenases to feed reducing equivalents into a single central pathway. Dehydrogenase activity is influenced by several factors including the rate at which the reductant is generated by intermediary metabolism, the affinity of the dehydrogenase for that reductant, the nature of its saturation kinetics (whether it is hyperbolic or sigmoidal) and its sensitivity to modulation by endogenous activators and inhibitors (e.g. NAD^+, AMP, oxaloacetate). In addition, a few dehydrogenases are regulated by the ambient energy charge (e.g. NADPH dehydrogenase in *A. vinelandii*). In contrast, cytochrome oxidase activity appears to be rate-limiting for respiration only under certain specialized growth conditions, such as during oxygen-limitation or in the presence of inhibitors (e.g. CN^-, CO). Under these conditions the rate of respiration reflects the ambient concentration of dissolved oxygen and the affinity of the oxidase for oxygen and/or inhibitor.

Additional to this type of kinetic regulation is the classical respiratory control exerted by the backpressure of the Δp as evidenced by the ability of suitable Δp-collapsing agents (uncouplers, various colicins, ADP plus phosphate) to stimulate respiration in carefully prepared suspensions of whole cells and/or membrane vesicles *in vitro*. The stimulation is generally fairly small, however, probably due to the presence of various energy-dissipating processes in whole cells (e.g. ion transport and various ATP-hydrolysing metabolic reactions) and the absence of tight energy coupling in imperfectly-sealed or heterogeneous membrane vesicles. Nevertheless, it is likely that this is the major control process *in vivo* when there is an excess of reductant and oxidant.

The major function of a branched respiratory system is probably to allow some flexibility in the exact pathway of electron transfer, thus enabling the organism to minimize the potentially deleterious effects of certain growth environments and to take maximum advantage of others. Thus, at low concentrations of oxygen, or in the presence of cyanide, electron transfer in a simple branched system such as *E. coli* is routed via the terminal branch which is most capable of maintaining a high respiratory rate. As a result, each oxidase carries that fraction of the total electron flux which reflects its concentration and kinetic properties relative to those of the other oxidase(s). Similarly, in the presence of excess oxygen, electron transfer is routed via the terminal branch which suffers least from the backpressure of the Δp. Thus, in a complex branched system in which the two terminal pathways and several major dehydrogenases are of unequal proton-translocating efficiency (see below) the overall respiratory route with the lowest efficiency will exhibit the fastest rate of electron transfer. This type of flexibility is of particular value to an organism such as *A. vinelandii* which needs to respire rapidly at high oxygen concentrations in order to protect an oxygen-sensitive nitrogen fixation apparatus, yet requires maximum energy conservation efficiency at other times.

The spatial organization of the respiratory chain

The spatial organization of the redox carriers within the membrane-bound respiratory chain is a particularly difficult problem to investigate since, unlike the ATP phosphohydrolase, they at most protrude only slightly from the surface of the membrane and hence can be visualized by electron microscopy only after staining with various relatively non-specific histochemical reagents (e.g. neotetrazolium). This problem has therefore principally been tackled by monitoring the interplay between membranes of opposed orientation (whole cells, protoplasts, sphaeroplasts and right-side-out vesicles compared with inside-out-vesicles) and membrane-impermeant agents. The latter include reductants (NADH, NADPH, ferrocytochrome c, ferrocyanide), oxidants NAD^+, $NADP^+$, ferricytochrome c, ferricyanide), activators and inhibitors (various nicotinamide and adenine nucleotides, antibodies, and weak acids such as cyanide and azide whose effectiveness is modified by the external pH and the polarity of Δp), functional group and surface labelling reagents (diazobenzene-[^{35}S]-sulphonate, Na[^{125}I] plus lactoperoxidase) and digestive enzymes (proteases, lipases). When added to right-side-out preparations, these agents react with redox carriers which are positioned on the surface of the membrane which originally faced the cell wall or periplasm, but not with those on the cytoplasmic surface; the converse is true with inside-out vesicles. This type of approach has been supplemented by measuring the distribution of paramagnetic redox centres and, in Gram negative organisms, by analysing the release of redox carriers into the periplasm following mild osmotic shock.

As yet, only relatively brief investigations have been carried out. The results indicate that the substrate-binding sites of transhydrogenase (for NADPH, $NADP^+$, NADH and NAD^+) and of malate, succinate, glycerol-3-phosphate, and D- and L-lactate dehydrogenases are on the cytoplasmic surface of the membrane, and the same appears to be true for the oxygen-binding sites of the cytochrome oxidases. In contrast, cytochrome c and methanol dehydrogenase are located on the surface of the coupling membrane which faces the cell wall or periplasm. Virtually nothing is known about the intramembrane positions of the iron-sulphur proteins or remaining cytochromes, but the hydrophobic nature of Q and MK probably ensures that they are embedded deeply within the membrane. Current knowledge of the organization and spatial distribution of redox carriers within the coupling membranes of aerobic chemoheterotrophs and facultative phototrophs is thus far from complete. The results so far are compatible with the chemiosmotic concept of

Bacterial Respiration and Photosynthesis

a transmembrane respiratory chain, but this fragmentary picture gives no real clue to the mechanism of respiration-linked proton translocation.

Respiration-linked proton translocation

There is now considerable evidence that the addition of a small pulse of dissolved oxygen, or an alternative electron acceptor, to a lightly-buffered anaerobic suspension of whole cells, protoplasts, sphaeroplasts or membrane vesicles prepared from a wide variety of chemoheterotrophs and facultative phototrophs initiates electrogenic proton translocation. The direction of H^+ movement is outwards with the various types of right-side-out preparations, and inwards with inside-out vesicles. When measured under conditions in which the concomitant movement of other ions is able to collapse the $\Delta\psi$ component of the protonmotive force (e.g. in the presence of K^+ plus the ionophorous antibiotic valinomycin, or of a permeant anion such as SCN^-) the observed rate of acidification by whole cells is extremely rapid and matches the rate of respiration (Fig. 2.3). The subsequent decay of the H^+ gradient is first order with respect to H^+ concentration and, with the exception of *Rhizobium leguminosarum* and some thermophilic bacteria, is relatively slow ($t\frac{1}{2} \geq 45\,s$). This slow rate, which reflects the inherently low permeability of the coupling membrane to H^+, is dramatically accelerated by lipophilic protonophores (e.g. carbonyl cyanide-*p*-trifluoromethoxyphenylhydrazone, FCCP). It should be noted, however, that when $\Delta\psi$ is not artificially collapsed, the rate of H^+ ejection is an order of magnitude lower than the rate of respiration. This might suggest, therefore, that the proton current is normally localized within the membrane, or very close to its surface, and only rapidly equilibrates with both of the adjacent aqueous phases (i.e. becomes chemiosmotic) under artificial conditions.

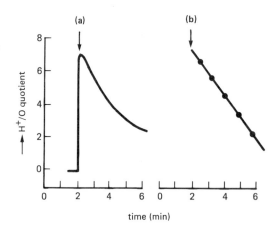

Fig. 2.3 Respiration-linked proton translocation initiated by the addition of a pulse of dissolved oxygen to a lightly-buffered anaerobic cell suspension. **(a)** linear plot, and **(b)** semi-logarithmic plot.

Aerobic respiration in chemoheterotrophs and facultative phototrophs

Such conditions also maximize the stoichiometry of respiration-linked proton translocation, and whole cell $\rightarrow H^+/O$ quotients of approximately 4 (e.g. *E. coli, K. pneumoniae* and *B. megaterium*), 6 (e.g. *M. lysodeikticus, P.* AM1 and *M. methylotrophus*) or 8 (e.g. *P. denitrificans, A. eutrophus, T. thermophilus* and *P. ovalis*) have been observed for the oxidation of endogenous substrates via respiratory chains of different redox carrier compositions. The number and location of the proton-translocating sites within these respiratory chains has been determined by measuring $\rightarrow H^+/O$ ($\rightarrow H^+/2e^-$) quotients after dissecting the chains into discrete functional segments through the use of selected reductants and oxidants of different redox potential. In this procedure, whole cells (or protoplasts and sphaeroplasts) are first starved of endogenous substrates, then loaded with exogenous substrates, both physiological (e.g. isocitrate, malate, lactate, succinate or methanol) or artificial (e.g. ubiquinol-1, duroquinol or ascorbate-TMPD) and finally pulsed with an appropriate oxidant (e.g. oxygen, Q-1, MK-O or ferricyanide). This technique is improved by adding inhibitors to block activity in unwanted regions of the respiratory chain (e.g. rotenone during the aerobic oxidation of flavin-linked substrates, and antimycin A or HQNO during the oxidation of cytochrome *c*-linked substrates). Such oxidant pulse experiments have recently been supplemented by initial rate experiments in which $\rightarrow H^+/O(\rightarrow H^+/2e^-)$ quotients are assayed from the initial rates of H^+ ejection

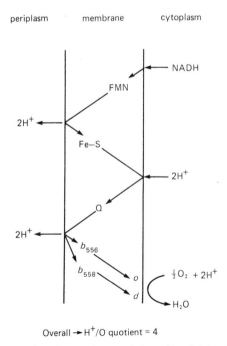

Overall $\rightarrow H^+/O$ quotient = 4

Fig. 2.4 The proton-translocating respiratory chain of *E. coli* (after Haddock and Jones, 1977). Note that cytochrome *d* is only present in significant amounts during oxygen-limited growth.

Bacterial Respiration and Photosynthesis

and oxygen reduction which follow the addition of various oxidizable substrates to lightly-buffered aerobic suspensions of cells in the presence of a suitable $\Delta\psi$-collapsing agent; →charge/O quotients (measured as →K^+/O quotients) have also been determined under similar conditions by measuring the initial rate of K^+ entry via added valinomycin. The results of these experiments, mainly of the oxygen pulse type, on chemoheterotrophic bacteria which exhibit different redox carrier patterns, in particular *E. coli, A. eutrophus, P. denitrificans, P.* AM1 and *M. methylotrophus* (including cytochrome-deficient mutants and phenotypically modified strains) have generally shown the presence of four proton- or charge-translocating sites, viz at the level of the energy-dependent transhydrogenase (site 0), NADH dehydrogenase (site 1) and the quinone-cytochrome system (sites 2 and 3). Sites 1 and 2 are generally constitutive (although site 1 is sometimes inoperative during iron- or sulphate-limited growth), whereas sites 0 and 3 are dependent on the variable abilities of different organisms to synthesize significant amounts respectively of an energy-linked transhydrogenase, and a high potential cytochrome *c* linked to cytochrome oxidases aa_3 or *o*. In those organisms which lack a cytochrome *c* of this type, respiration is terminated by cytochrome oxidases aa_3, *o* or *d* immediately after the Q(MK)-*b* region.

In chemiosmotic terms, each proton translocating respiratory segment was originally envisaged as a transmembrane redox loop consisting of a hydrogen carrier (e.g. flavin, quinone) followed by an electron carrier (e.g. Fe-S, cytochrome). The ejection of $2H^+$ would thus occur via the outward transfer of 2H followed by the inward transfer of $2e^-$, giving an →H^+/site ratio of 2 (see Fig. 1.4). The composition and possible sequences of the redox carriers which comprise sites 1 and 2 in *E. coli* and other organisms with similar respiratory chains are compatible with such a redox loop mechanism, and the little that is known about the spatial organization of these carriers within the respiratory membrane does not conflict with this view (Fig. 2.4). In contrast, organisms such as *A. eutrophus* and *M. methylotrophus*, which contain cytochrome *c* and exhibit higher →H^+/O quotients than *E. coli*, appear to lack the hydrogen carrier that is required for proton translocation at site 3 via a redox loop mechanism. However, it is possible that in these organisms site 2 is organised into a complicated protonmotive quinone cycle which involves the quinone, semiquinone and quinol forms of Q or MK, and which catalyses the ejection of $4H^+ (2 \times 2H^+)$ during two successive one-electron transfers from an Fe-S centre in the primary dehydrogenase to cytochrome *c* (→H^+/O = 4, →K^+/O = 2); respiration is completed by the inward transfer of electrons from *c* to molecular oxygen via cytochrome oxidases aa_3 or *o*, but not *d* (→H^+/O = 0, →K^+/O = 2) (Fig. 2.5). Although there is relatively little direct evidence to support the concept of a protonmotive cycle, vectorial electron transfer via the terminal cytochrome system is fully supported by the observed →H^+/O and →K^+/O quotients for the oxidation of the artificial cytochrome-linked reductant ascorbate-TMPD, and is compatible with the known spatial organization of the redox carriers within the membrane. An electron-transferring redox arm of this type would also suffice to translocate protons at site 3 during the oxidation of methanol (→H^+/O = 2, →K^+/O = 2), since the proton-releasing methanol dehydrogenase is located on the periplasmic side of the membrane.

Rather interestingly, higher than expected efficiencies of respiration-linked proton translocation have been reported for the oxidation of $NAD(P)^+$-linked substrates by *P. denitrificans* (→H^+/O up to 10) and the moderate thermophile *Bacillus stearothermophilus* (→H^+/O = 8, →K^+/O = 8; no transhydrogenase

Aerobic respiration in chemoheterotrophs and facultative phototrophs

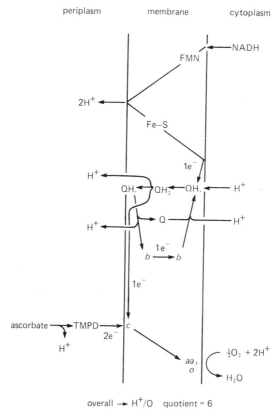

Fig. 2.5 The proton-translocating respiratory chain of organisms that contain cytochrome c. Note that the protonmotive quinone cycle is largely unproven experimentally, and that in *P. denitrificans* cytochrome oxidase aa_3 may act as a proton pump.

present); in addition, the former organism exhibits an $\rightarrow H^+/O$ quotient of 4 for the oxidation of methanol via the terminal $c \rightarrow aa_3$ pathway. It is likely, therefore, that both of these organisms contain a cytochrome oxidase aa_3 which effectively translocates $2H^+$ and four charges per electron pair transferred, i.e. like the corresponding enzyme from mitochondria it acts as a proton pump. Thus it is possible that there are two types of cytochrome oxidase aa_3 in bacteria, viz. those which catalyse only electron transfer ($\rightarrow H^+/O = 0, \rightarrow K^+/O = 2$) and those which catalyse both electron transfer and proton translocation ($\rightarrow H^+/O = 2, \rightarrow K^+/O = 4$).

It must be concluded, therefore, that the redox loop/arm mechanism of proton translocation in bacteria has not been disproved (at least at site 1 and, with certain substrates, at site 3), although the redox cycle mechanism for site 2 is much more speculative. Indeed, it is possible that proton translocation in the quinone-*b* region, and also at the level of cytochrome oxidase aa_3 in some organisms, occurs

Table 2.3 Methods of detecting energy coupling sites in bacterial respiratory chains

Assay material	Method of assay
In vivo cultures	Molar growth yields ($Y_{O_2}^{max}$, $Y_{carbon\ substrate}^{max}$).
Whole cells, sphaeroplasts, protoplasts and right-side-out vesicles	$\rightarrow H^+/O (\rightarrow H^+/2e^-)$ quotients.
	ATP/O (ATP/$2e^-$) quotients.
	Ion uptake (e.g. of amino acids, TPMP$^+$ or ^{86}Rb$^+$) at the expense of forward electron transfer over defined segments of the respiratory chain.
Inside-out vesicles	ATP/O (ATP/$2e^-$) quotients.
	ADP/O quotients from respiratory control cycles.
	Reversed electron transfer (e.g. NADP$^+ \leftarrow$ NADH, NAD$^+ \leftarrow$ succinate, cytochrome $b \leftarrow c$) at the expense of
	(i) ATP hydrolysis
	(ii) forward electron transfer through sites 1, 2 or 3.
	Ion uptake (e.g. of SCN$^-$ or ANS$^-$) at the expense of forward electron transfer over defined segments of the respiratory chain.
	Stimulation of forward electron transfer by uncoupling agents.

$Y_{O_2}^{max}$, maximum molar growth yield with respect to oxygen consumption (g cells.mole O_2^{-1}); $Y_{carbon\ substrate}^{max}$, maximum molar growth yield with respect to carbon substrate utilization (e.g. g cells.mole glucose^{-1}); TPMP$^+$, triphenylmethylphosphonium; ANS$^-$, anilinonaphthalene sulphonate.

via, a proton pump (membrane Bohr effect). Furthermore, the absence of redox carriers from the energy-linked transhydrogenase suggests that the latter mechanism is also responsible for H^+ translocation at site 0.

Energy coupling sites

The organization of bacterial respiratory chains into proton-translocating segments is of course compatible with the classical concept of separate energy coupling sites located within well defined regions of the chain (it does not necessarily mean, however, that each site synthesizes exactly one molecule of ATP per $2e^-$ transferred). The location of these energy coupling sites via the direct assay of ATP synthesis during respiration is considerably more difficult in bacteria than in mitochondria, since the absence of an adenine nucleotide translocase from the coupling membranes of free-living bacteria renders whole cells incapable either of utilizing exogenous ADP as a phosphoryl acceptor for oxidative phosphorylation or of catalyzing the hydrolysis of exogenous ATP. Furthermore, attempts to measure ATP/O (ATP/$2e^-$) quotients in whole cells, by monitoring changes in the intracellular concentrations of adenine nucleotides following the initiation of respiration by small additions of dissolved oxygen to anaerobic cells, or of oxidizable substrate to aerobic cells, have yielded variable results (mainly because of the difficulties of assaying intracellular ATP synthesis net of any competing hydrolytic reactions).

The detection and location of energy coupling sites in bacteria has therefore traditionally been carried out using inside-out membrane vesicles via the direct assays of ATP/O (or ATP/$2e^-$) quotients or, to a lesser extent, ADP/O quotients from respiratory control cycles. However, since these vesicles are rarely either perfectly homogeneous or completely sealed, energy transduction efficiencies have been generally low (maximum ATP/O quotients of 1.0 to 1.5 for the oxidation of NADH); quantitative measurements are thus of limited use in determining the number of energy coupling sites. A more qualitative approach has therefore been adopted in which those segments of the respiratory chain are identified which (i) exhibit uncoupler-stimulated respiration, (ii) drive ATP synthesis or ion transport at the expense of forward respiration, and (iii) exhibit reversed respiration at the expense either of ATP hydrolysis or of forward respiration through a different energy coupling site (Table 2.3).

Using these approaches, classical energy coupling sites 1 and 2 have been detected in membrane vesicles from a wide range of bacteria which exhibit respiration-linked proton translocation at these two sites: these include *E. coli, Pseudomonas denitrificans, M. lysodeikticus, M. phlei, P.* AM1 and *A. vinelandii*. Interestingly, site 1 appears to be missing from the thermophile *B. caldolyticus*, which additionally has the peculiar property of requiring energy for succinate oxidation; site 1 is also absent from nitrogen-fixing cultures of *A. vinelindii* cultured under highly aerobic conditions, and can be variably deleted from *E. coli* and *P. denitrificans* by growing these organisms under conditions of sulphate- or iron-limitation. ATP synthesis at site 3 depends on the presence of cytochrome *c* linked to cytochrome oxidases aa_3 or *o* (e.g. in *P. denitrificans, P.* AM1, *M. lysodeikticus, M. phlei, B. caldolyticus* and the minor terminal pathway of *A. vinelandii*). This has been neatly confirmed in *P.* AM1 by showing that cytochrome *c* deficient mutants fail not only to oxidize methanol but also to catalyse energy coupling at site 3, and in

Bacterial Respiration and Photosynthesis

P. denitrificans by demonstrating that right-side-out vesicles readily catalyse solute uptake during ascorbate-TMPD oxidation. Energy coupling at site 0 is dependent upon the presence of an energy-linked transhydrogenase (e.g. *E. coli*, *A. eutrophus* and *Pseudomonas denitrificans*). In contrast to the other three coupling sites, which exhibit $\Delta E_0'$ values exceeding 200 mV, the $\Delta E_0'$ value of the transhydrogenase reaction is only about 4 mV. Thus, since the Δp generated by an individual proton-translocating site is proportional to the difference in redox potential across that site, the transhydrogenase reaction will only generate a Δp sufficient to drive ATP synthesis when the [NADPH][NAD$^+$]/[NADP$^+$][NADH] ratio is very high so that the redox potential difference approaches 200 mV. Since these nicotinamide nucleotide concentrations are unlikely to pertain *in vivo*, site 0 is probably of little significance to oxidative phosphorylation and it is much more likely that it catalyses the energy-dependent reduction of NADP$^+$ by NADH at the expense of energy conserved during forward respiration through the other energy coupling sites.

These *in vitro* studies of respiratory chain energy conservation have been supplemented by *in vivo* determinations of the maximum molar growth yields of energy-limited continuous cultures (i.e. the amount of cells produced, in grams per mole of oxygen reduced or carbon source utilised, when corrected, as far as is possible, for the expenditure of energy for maintence purposes; $Y_{O_2}^{max}$, $Y_{carbon\ substrate}^{max}$). The results of comparative studies generally indicate that those organisms which route the majority of their terminal electron flow via a cytochrome $c \rightarrow aa_3/o$ system exhibit molar growth yields which are approximately 50% higher than those which lack cytochrome c. The former organisms contain three equivalent energy conservation sites (1, 2 and 3), whereas the latter lack site 3; the presence of site 0 does not significantly affect the yields. It is interesting to note, however, that *R. leguminosarum* and the thermophiles *B. stearothermophilus* and *T. thermophilus* exhibit particularly low yields compared with mesophiles of similar respiratory chain composition. It would appear, therefore, that these organisms waste a considerable proportion of their conserved energy, possibly via enhanced H$^+$ leakage through the coupling membrane.

The protonmotive force (Δp)

There is now substantial evidence that aerobic respiration in chemoheterotrophs and facultative phototrophs generates a transmembrane Δp *in vitro* which is variably comprised of ΔpH and $\Delta \psi$ (internal compartment alkaline and electrically negative in whole cells, protoplasts, sphaeroplasts and right-side-out vesicles, but acidic and electrically positive in inside-out vesicles). Δp has recently been quantitated for several species of neutrophilic bacteria, in particular *E. coli*, *P. denitrificans*, *Staphylococcus aureus*, *M. methylotrophus* and *T. thermophilus*.

The most frequently used method involves measuring ΔpH and $\Delta \psi$ from the transmembrane distribution of permeant marker molecules during respiration. Since weak acids and bases penetrate membranes much more readily in their neutral forms (HA, B) than in their ionised states (A$^-$, BH$^+$), their concentrations on either side of the coupling membrane following energization reflect the ΔpH; weak acids will therefore accumulate in the alkaline compartment and the weak bases in the acidic compartment. The ΔpH of whole cells, sphaeroplasts and right-side-out vesicles is thus routinely determined from the distribution of a weak acid (e.g. [2-

^{14}C]-5, 5-dimethyloxazolidine-2, 4-dione, DMO), whereas a weak base (e.g. [^{14}C]-methylamine or the fluorescent probe 9-aminoacridine) is a much more effective indicator of the ΔpH of inside-out vesicles. For the determination of $\Delta\psi$, lipophilic organic and inorganic ions are used which freely permeate the coupling membrane and which therefore distribute themselves according to $\Delta\psi$. A suitable cation (e.g. [^{3}H]-triphenylmethylphosphonium, TPMP$^+$, or ^{86}Rb$^+$ in the presence of valinomycin) is used to determine $\Delta\psi$ in whole cells, sphaeroplasts or right-side-out vesicles, whereas an anion (e.g. [^{14}C]-SCN$^-$ or the fluorescent probe 8-anilinonaphthalene-1-sulphonate, ANS$^-$) is more applicable to inside-out vesicles. Ion distributions can also be monitored directly using appropriate electrodes, and microelectrodes have been used to measure $\Delta\psi$ in giant cells of *E. coli* produced by growth in 6-amidino-penicillanic acid. Finally, ΔpH has been quantitated by high resolution ^{31}P nuclear magnetic resonance spectroscopy.

The results of these various approaches indicate Δp values within the range 120 to 210 mV for the oxidation of NADH, succinate, methanol or ascorbate-TMPD by whole cells, sphaeroplasts and/or membrane vesicles (the latter being analysed in the absence of any competing reactions such as ATP synthesis or reversed respiration, i.e. under so-called 'static-head' conditions). Extensive studies with *E. coli* have shown that ΔpH and $\Delta\psi$ contribute variably to Δp, depending on the pH on the periplasmic side of the membrane. Thus, ΔpH is maximal at pH 5.5 (approximately 2 units; equivalent to 120 mV) but falls to zero at pH 7.5 and becomes reversed at more alkaline pH values; $\Delta\psi$ undergoes converse changes to maintain Δp essentially constant at approximately 160 to 180 mV. The mechanism via which this organism maintains the pH on the cytoplasmic side of the membrane (i.e. the internal pH in whole cells) at approximately 7.5 is probably extremely complicated and may involve the transport of Na$^+$ and K$^+$.

The composition of the transmembrane Δp can also be dramatically altered by the addition of specific ΔpH- or $\Delta\psi$-collapsing agents (Fig. 2.6). Thus ΔpH is eliminated by the ionophorous antibiotic nigericin + K$^+$, and $\Delta\psi$ is eliminated by SCN$^-$ or by valinomycin + K$^+$. Under steady state conditions:

$$\Delta p = J_{H^+}/Cm_{H^+}$$

Where J_{H^+} is the rate of proton translocation (nmol H$^+$.min^{-1}.mg) and Cm_{H^+} is the effective proton conductance of the coupling membrane (nmol H$^+$.min^{-1}.mg^{-1}.mV^{-1}). Since the rate of proton backflow is proportional to Δp, the dissipation of either ΔpH or $\Delta\psi$ allows the other to increase and hence maintain Δp essentially constant, i.e. the collapse of ΔpH is accompanied by the enhancement of $\Delta\psi$, and *vice-versa*. Both components of Δp are collapsed by FCCP, valinomycin + nigericin + K$^+$, and NH$_4^+$ salts (although the latter are only effective with inside-out vesicles because they must be added to the alkaline compartment). Since only these last three reagents will significantly inhibit oxidative phosphorylation and reversed electron transfer, it is clear that ΔpH and $\Delta\psi$ are essentially interchangeable as far as these two processes are concerned, and that uncoupling only occurs when both components of Δp are dissipated. However, the transport of some solutes across the coupling membrane is specifically dependent on ΔpH or $\Delta\psi$; in the former case transport is specifically inhibited by nigericin + K$^+$, and in the latter case by valinomycin + K$^+$.

The wide variations in the reported values of Δp (120 to 210 mV approximately) probably reflect differences in the source, redox carrier composition and integrity of the various membrane preparations used, as well as differences in the nature of the

Bacterial Respiration and Photosynthesis

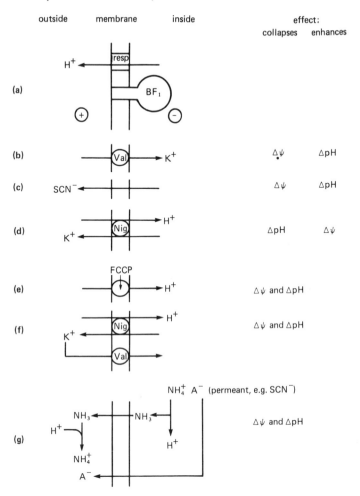

Fig. 2.6 The effect of various reagents on the composition of the Δp generated by aerobic respiration. (a) no additions, (b) valinomycin + K^+, (c) SCN^-, (d) nigericin + K^+, (e) FCCP, (f) valinomycin + nigericin + K^+, and (g) NH_4^+ + a permeant anion. Note that the latter combination is only effective when added to inside-out vesicles (i.e. when it can reach the alkaline and electrically negative compartment); the other reagents are independent of the orientation of the coupling membrane. Only reactions (e) to (g) uncouple ATP synthesis. Valinomycin and nigericin act as mobile ion carriers (V.K^+, N^-K^+, N^-H^+), although for simplicity they are shown here as forming channels.

Aerobic respiration in chemoheterotrophs and facultative phototrophs

oxidizable substrate and imperfections in some or all of the assay procedures employed. In spite of this, there is little doubt that aerobic respiration in neutrophilic chemoheterotrophs can generate protonmotive forces of the correct sign and of the required magnitude (Chapter 5) to drive ATP synthesis and other energy-dependent membrane reactions. Whether such delocalized transmembrane forces are actually involved in the mechanism of energy transduction *in vivo*, however, remains to be determined.

Bacteria which live at extremes of pH, i.e. the acidophiles and alkaliphiles, pose intriguing problems as far as a chemiosmotic mechanism of energy transduction is concerned. There is convincing evidence that these organisms do not contain enzymes with appropriately low or high pH optima, and that they maintain their cytoplasm at a pH significantly closer to neutral than the external environment, i.e. approximately 6 in acidophiles, and close to 9 in alkaliphiles. Thus, for example, *B. acidocaldarius* growing at pH 2 maintains a ΔpH of approximately 4 units ($\equiv$ 240 mV; inside alkaline). In order to carry out steady state respiration against such a high back pressure, one or more respiratory adaptations are required as dictated by the equation:

$$\rightarrow H^+/O . \Delta p = 2\Delta E_h$$

where ΔE_h is the actual (non-standard) redox potential difference between the initial and final substrates. Thus, either the stoichiometry of respiration-linked proton translocation must be low and/or ΔE_h must be high (interestingly, *B. acidocaldarius* lacks cytochrome *c* and yields a maximal $\rightarrow H^+/O$ quotient of approximately 4). Alternatively, or additionally, the very high ΔpH must be offset by a reversed $\Delta\psi$, such that Δp is lower than ΔpH (indeed, a reversed $\Delta\psi$ of approximately 34 mV, positive inside, has been detected with whole cells of *B. acidocaldarius*, which thus decreases Δp to approximately 200 mV). The magnitude and direction of the latter are therefore compatible with a chemiosmotic mechanism of ATP synthesis, but the method by which the reversed $\Delta\psi$ is generated is not currently known.

In contrast, alkaliphiles such as *B. alcalophilus* or *B. pasteurii* growing at pH 10 maintain a reversed ΔpH of approximately 1 unit ($\equiv$ 60 mV) which is probably effected via a highly active H^+. Na^+ antiport, the latter being absent from a non-alkaliphilic mutant of *B. alcalophilus*. In order to obtain the requisite Δp for ATP synthesis via chemiosmosis, these organisms would need to generate a $\Delta\psi$ of up to 260 mV. Although both organisms generate very significant membrane potentials, values of this magnitude have not been detected; indeed, they are probably incompatible with the maintenance of normal membrane behaviour, and possibly also with transmembrane electron flow. The overall Δp of *B. alcalophilus*, for example, does not exceed approximately 84 mV and is therefore quite inadequate to drive ATP synthesis by a chemiosmotic mechanism unless the $\rightarrow H^+/ATP$ quotient is particularly high (Chapter 5). It is possible, therefore, that in these alkaliphiles Δp does not reflect the true driving force for energy transduction, and that the latter occurs via a more localized mechanism.

The biased composition of Δp in these organisms (predominantly ΔpH in acidophiles, and $\Delta\psi$ in alkaliphiles) has interesting implications for solute transport. Indeed, there is some evidence that the range of growth substrates utilized by acidophiles is restricted to those which normally enter the cell via electroneutral H^+ symport, whereas alkaliphiles predominantly use substrates which enter either in response to $\Delta\psi$ (directly or in cotransport with Na^+) or via ATP hydrolysis. The low Δp in alkaliphiles does not preclude solute transport via a

Summary

The aerobic respiratory chains of chemoheterotrophic and facultatively phototrophic bacteria contain the same basic types of redox carriers as those present in higher organisms. Considerable interspecies diversity is apparent, however, particularly in the identities of the quinones and cytochrome oxidases, and in the presence or absence of transhydrogenase and cytochrome c. A variety of experimental approaches, including phenotypic and genotypic manipulations, have shown that the redox carriers are sequentially organized into respiratory pathways of varying complexity, particularly with respect to the terminal cytochrome system which can be either linear or branched; branching often occurs in response to environmental conditions. There is also evidence that the redox carriers are spatially organized within the membrane, such that although respiration usually starts and finishes on the cytoplasmic surface it stretches across to the external surface, possibly more than once.

All of the respiratory chains so far examined exhibit respiration-linked proton translocation. Depending on their redox carrier compositions they contain a variable number of proton or charge-translocating segments, each of which exhibits an $\rightarrow H^+$/site ratio ≥ 0 and a $\rightarrow K^+$/site ratio ≥ 2, and is equivalent to a classical energy coupling site. Sites 1 and 2 are almost invariably present, but sites 0 and 3 are dependent upon the presence respectively of an energy-linked transhydrogenase and cytochrome c linked to cytochrome oxidases aa_3 or o. Sites 1, 2 and 3 generate a transmembrane Δp of up to approximately 200 mV, and hence are capable of driving ATP synthesis and other energy-dependent membrane functions. In contrast, site 0 does not contribute to oxidative phosphorylation *in vivo* and probably serves to catalyse energy-dependent transhydrogenation ($NADP^+ \leftarrow NADH$) at the expense of Δp generated at the other sites.

The protonmotive force generated by respiration is variably composed of ΔpH and $\Delta \psi$, the relative contributions of which can be varied by the addition of specific ΔpH- and $\Delta \psi$-collapsing agents. Uncoupling of ATP synthesis occurs only when both components of Δp are dissipated, but some transport reactions are inhibited following the selective abolition of ΔpH or $\Delta \psi$. The composition of Δp also varies as a function of the environmental pH, ΔpH and $\Delta \psi$ predominating under acidic and alkaline conditions respectively. Such variations enable the intracellular pH to be kept relatively constant. Most of the experimental evidence is compatible with a chemiosmotic mechanism of energy transduction, although some observations can be interpreted as favouring a more localized proton current.

References

HADDOCK, B. A. (1977). The isolation of phenotypic and genotypic variants for the functional characterisation of bacterial oxidative phosphorylation. In: *Microbial Energetics* pp. 96–120. Edited by B. A. Haddock and W. A. Hamilton. Society for General Microbiology Symposium 27. Cambridge University Press, Cambridge.

HADDOCK, B. A. and JONES, C. W. (1977). Bacterial respiration. *Bacteriological Reviews* 41: 47–99.

Aerobic respiration in chemoheterotrophs and facultative phototrophs

JOHN, P. and WHATLEY, F. R. (1977). The bioenergetics of *Paracoccus denitrificans*. *Biochimica et Biophysica Acta* 463: 129–53.

JONES, C. W. (1977). Aerobic respiratory systems in bacteria. In: *Microbial Energetics* pp. 23–59. Edited by B. A. Haddock and W. A. Hamilton. Society for General Microbiology Symposium 27. Cambridge University Press, Cambridge.

KNOWLES, C. J. (1980). Ed. *Diversity of Bacterial Respiratory Systems* Vol 1. CRC Press, Boca Raton, Florida.

LUDWIG, B. (1980). Haem aa_3-type cytochrome c oxidases from bacteria. *Biochimica et Biophysica Acta* 594: 177–189.

STOUTHAMER, A. H. (1980). Bioenergetic studies on *Paracoccus denitrificans*. *Trends in Biochemical Sciences* 5: 164–6.

TEMPEST, D. W. (1978). The biochemical significance of microbial growth yields: a reassessment. *Trends in Biochemical Sciences* 3: 180–4.

3 Aerobic respiration in chemolithotrophs; anaerobic respiration

Many species of bacteria contain specialized respiratory chains that oxidize inorganic reductants instead of reduced carbon compounds, or catalyse the dissimilatory reduction of various inorganic or organic oxidants as alternatives to molecular oxygen. The first group of bacteria, the chemolithotrophs or chemoautotrophs, are relatively restricted taxonomically and are comprised of only three major families plus a few tenuously related genera (Table 3.1). The majority of these organisms are obligate aerobes, a few are either thermophilic or acidophilic, and some are obligate chemoautotrophs, i.e. they obtain their cell carbon by reductively assimilating carbon dioxide at the expense of NAD(P)H. The reductants used by chemolithotrophs include relatively reduced nitrogen and sulphur compounds, hydrogen, ferrous iron and, paradoxically, carbon monoxide. Since the majority of these compounds have redox potentials which are too high to reduce $NAD(P)^+$ directly, a variable proportion of the energy which is conserved during respiration is used to drive the formation of NAD(P)H via reversed electron transfer. In contrast, those mainly chemoheterotrophic organisms which are capable of catalysing anaerobic respiration are more widely distributed taxonomically and include both facultative and obligate anaerobes, the majority of which are Gram negative mesophiles (Table 3.2). The oxidants used by these organisms include relatively oxidized nitrogen and sulphur compounds, fumarate, carbon dioxide, ferric iron and trimethylamine-N-oxide, (TMAO), most of which are relatively weak oxidants compared with oxygen. The oxidation and reduction of nitrogen, sulphur and iron compounds by these two groups of bacteria *in vivo* are largely responsible for the nitrogen, sulphur and iron cycles, respectively.

The oxidation and reduction of nitrogen compounds

The nitrogen cycle The cyclic oxidation and reduction of nitrogen compounds occurs via consecutive one- or two-electron transfers from ammonia (oxidation state -3) to nitrate ($+5$) and back again (Fig. 3.1). The oxidative, nitrifying half of the cycle is an aerobic sequence of reactions, whereas the reductive denitrifying half is essentially anaerobic although it includes nitrogen fixation which can occur, albeit to a limited extent, in aerobic bacteria. Free energy is released during both nitrification and denitrification but the latter, which principally uses NADH as reductant, generally transfers electrons over a much wider redox potential range (Table 3.3) and hence is associated with higher ATP yields.

The oxidation of ammonia to nitrite This series of reactions is carried out almost solely by soil organisms of the genus *Nitrosomonas* and related nitrifying bacteria, and occurs via the formation of hydroxylamine (-1) and probably nitroxyl (NOH; $+1$). The E'_θ value of the NH_2OH/NH_3 couple ($+899$ mV) is too high to allow

Table 3.1 Bacteria which are capable of obtaining energy from the oxidation of inorganic electron donors (Chemolithotrophs)

Family	Genus	Electron donor	Notes
Nitrobacteraceae	Nitrosomonas and three other genera Nitrobacter and two other genera	$NH_3 (\rightarrow NO_2^-)$ $NO_2^- (\rightarrow NO_3^-)$	*Nitrifying bacteria*. Obligately aerobic and chemolithotrophic (except for *Nitrobacter winogradskyi* which also grows heterotrophically). Obligately autotrophic
Thiobacillaceae	Thiobacillus Sulpholobus	$S^{2-}, S°, S_2O_3^{2-}, SO_3^{2-} (\rightarrow SO_4^{2-})$	*Sulphur oxidizing bacteria.* *Thiobacillus*. Obligately aerobic (except for *T. denitrificans* which reduces NO_3^-). Obligately chemolithotrophic and autotrophic (except for *T. novellus* which is facultatively, autotrophic). *T. ferro-oxidans* also oxidizes $Fe^{2+} (\rightarrow Fe^{3+})$ *Sulpholobus*. Obligately aerobic and chemolithotrophic; facultatively autotrophic. Acidophilic and thermophilic.
	Thermothrix and three other genera		*Thermothrix*. Facultatively anaerobic (reduces O_2, NO_2^-, NO_3^-), chemolithotrophic and autotrophic. Thermophilic.
Thiobacillaceae	Thiobacillus	$Fe^{2+} (\rightarrow Fe^{3+})$	*Iron bacteria*. Little is known about these organism and only *T. ferro-oxidans* has been studied in detail.
—	Alcaligenes (Hydrogenomonas)	H_2	*Hydrogen bacteria*. Selected species only of these genera; some species will replace oxygen with alternative oxidants. Obligately aerobic or facultatively anaerobic.
—	Paracoccus		Facultatively chemolithotrophic.
Pseudomonadaceae	Pseudomonas plus several other genera, including sulphate-reducers and methanogens.		

Bacterial Respiration and Photosynthesis

Table 3.2 The major genera of bacteria which are capable of obtaining energy via anaerobic respiration

Family	Genus	Electron acceptor	Notes
Enterobacteriaceae	Escherichia, Klebsiella and all other genera	$NO_3^- (\to NO_2^-)$, $NO_2^- (\to NH_3)$	Facultatively anaerobic.
Bacillaceae Pseudomonadaceae	Bacillus Pseudomonas	$NO_3^- (\to NO_2^-)$, $NO_2^- (\to N_2O, N_2)$	*Denitrifying bacteria.* Facultatively anaerobic. Some species only of *Bacillus* (e.g. *B. licheniformis*) and *Pseudomonas* (e.g. *P. aeruginosa*, *P. stutzeri*) are capable of denitrification.
— —	Alcaligenes Paracoccus		
Streptomycetaceae	Streptomyces	$NO_2^- (\to NO?)$	
Vibrionaceae	Vibrio	fumarate ($\to$ succinate)	Obligately anaerobic or facultatively anaerobic. Not all species of every genus are capable of fumarate reduction.
Enterobacteriaceae — Bacteroidaceae Streptococcaceae Bacillaceae Propionibacteriaceae Bacillaceae	Escherichia, Klebsiella and all other genera Haemophilus Bacteroides Streptococcus Bacillus Propionibacterium Desulfovibrio		
Bacillaceae	Desulfotomaculum Desulfovibrio Desulfuromonas	$SO_4^{2-} (\to APS^{2-} \to SO_3^{2-} \to S^{2-})$ $S° (\to \to \to S^{2-})$	*Sulphate-reducing bacteria.* Obligately anaerobic. Some species of *Desulfotomaculum* are moderately thermophilic, some species of *Desulfovibrio* are moderately psychrophilic or halophilic.

Aerobic respiration in chemolithotrophs; anaerobic respiration

Methanobacteriaceae	Methanobacterium and two other genera	$CO_2(\to CH_4)$	Methane-producing bacteria. Obligately anaerobic. Some species of Methanobacterium are thermophilic.
Bacillaceae Pseudomonadaceae	Bacillus Pseudomonas	$Fe^{3+}(\to Fe^{2+})$	Facultatively anaerobic; selected species only. Acid tolerant.
Enterobacteriaceae	Escherichia, Proteus and several other genera	TMAO($\to$TMA)	Facultatively anaerobic. Some species of Pseudomonas and Alteromonas are psychrophilic.
Pseudomonadaceae Vibrionaceae	Pseudomonas Alteromonas		

Table 3.3 The E'_0 values of redox couples involved in nitrification and denitrification

Nitrification		Denitrification	
Redox couple	E'_0(mV)	Redox couple	E'_0(mV)
NO_2^-/NH_2OH	+66	$NAD^+/NADH$	−320
NO_3^-/NO_2^-	+420	NO_2^-/NO	+374
$\frac{1}{2}O_2/H_2O$	+820	NO_3^-/NO_2^-	+420
(NH_2OH/NH_3)	+899		
		NO/N_2O	+1175
		N_2O/N_2	+1355

Bacterial Respiration and Photosynthesis

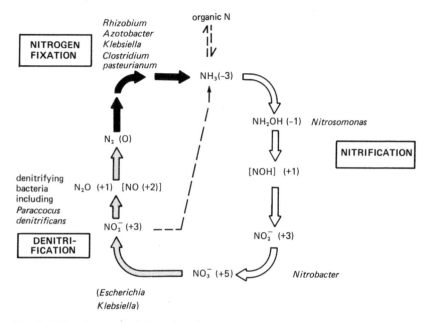

Fig. 3.1 The nitrogen cycle. The dotted arrows represent the reductive assimilation and subsequent decomposition of organic nitrogen compounds. The dashed arrows indicate a short-circuit via which nitrite is reduced to ammonia in *E. coli K. pneumoniae* and other facultative anaerobes, thereby avoiding nitrogen fixation which is the rate-limiting step of the cycle.

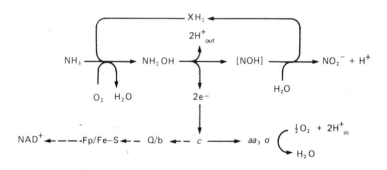

Overall: $NH_3 + 1\tfrac{1}{2} O_2 \longrightarrow NO_2^- + H_2O + H^+$

Fig. 3.2 The oxidation of ammonia to nitrite by *Nitrosomonas*. The dashed line indicates energy-dependent reversed respiration. Note that the cytochrome P460-dependent reactions occur in the cytoplasm.

Aerobic respiration in chemolithotrophs; anaerobic respiration

ammonia plus water to be oxidized to hydroxylamine via the respiratory chain

$$\left.\begin{array}{c} NH_3 + H_2O \\ NH_2OH \end{array}\right\} \xrightarrow{[2H]} \left(\begin{array}{c} \frac{1}{2}O_2 \\ H_2O \end{array}\right.$$

and it is much more likely that the formation of hydroxylamine is accomplished via an oxygenation reaction

$$NH_3 + XH_2 + O_2 \longrightarrow NH_2OH + X + H_2O$$

involving ammonia hydroxylase and cytochrome P460, a specialized b-type cytochrome (Fig. 3.2). It is possible that XH_2 can be equated with reduced $P460 + 2H^+$, but the exact mechanism of the reaction is currently unclear. Interestingly, it bears some similarity to the oxygenation of methane to methanol by methanotrophic bacteria ($CH_4 + XH_2 + O_2 \rightarrow CH_3OH + H_2O + X$), a reaction which is catalysed by an oligomeric iron-sulphur protein, methane monooxygenase, and in which XH_2 is NADH or possibly reduced cytochrome c according to species. However, although many methanotrophs will oxygenate ammonia to hydroxylamine, nitrifying bacteria cannot convert methane to methanol.

The subsequent oxidation of hydroxylamine to nitroxyl [NOH] by the genus *Nitrosomonas* is catalysed by hydroxylamine-cytochrome c reductase and the terminal cytochrome system and rather unexpectedly exhibits an $\rightarrow H^+/O$ quotient of only 2. However, since this organism contains an extremely complex intracytoplasmic membrane system composed of concentric layers, it is possible that this $\rightarrow H^+/O$ quotient has been underestimated; indeed, it is difficult to envisage how an entirely delocalized proton current is responsible for energy transduction under these circumstances. Cell-free preparations of members of the genus *Nitrosomonas* are poorly coupled in terms of ATP synthesis, but will nevertheless catalyse the reduction of NAD^+ by hydroxylamine or ferrocytochrome c (sites 1 and 2) at the expense of energy released by ATP hydrolysis or aerobic respiration (site 3). Nitroxyl is highly unstable and is rapidly converted to nitrite via a second cytochrome P460-dependent oxygenation reaction.

Virtually nothing is known about the spatial organization of the respiratory chain in species of *Nitrosomonas*, but since hydroxylamine is generated in the cytoplasm it is likely either that hydroxylamine-cytochrome c reductase and cytochrome oxidase constitute a proton-translocating redox loop, or that respiration is scalar and cytochrome oxidase pumps $2H^+$. Since the overall oxidation of ammonia to nitrite consumes $1.5\ O_2$ and probably produces only 1 ATP equivalent, some of which is used to drive reversed electron transfer and carbon dioxide assimilation, the molar growth yield of this organism is extremely low.

The oxidation of nitrite to nitrate This reaction is carried out mainly by another genus of obligately aerobic soil organisms, *Nitrobacter*. Like the oxidation of ammonia to hydroxylamine, it can theoretically occur via either oxygenation or respiration. In this case, however, the E'_θ value of the nitrate/nitrite couple ($+420\,mV$) is compatible with a respiratory mechanism, and this has been confirmed by studies with ^{18}O which show that nitrite receives an atom of oxygen

Bacterial Respiration and Photosynthesis

from water rather than from atmospheric oxygen:

$$H_2O + NO_2^- \underset{NO_3^-}{\Big)} \xrightarrow{[2H]} \underset{H_2O}{\Big(} \tfrac{1}{2}O_2$$

Careful investigations of the respiratory chains of *N. winogradskyi* and, to a lesser extent, *N. agilis* have indicated that they contain a plethora of high redox potential components which include two non-autoxidizable cytochromes a_1 ($E_m = +140$ and $+350$ mV), cytochrome c ($E_m = +270$ mV), cytochrome oxidase aa_3 ($E_m = +240$ and $+400$ mV), several [2Fe-2S] and [4Fe-4S] proteins with E_m values in the range $+60$ to $+320$ mV, and a molybdenum centre ($E_m = +340$ mV). The coupling membrane is organized into intracytoplasmic invaginations, similar to mitochondrial cristae, from which inside-out membrane vesicles with unusually high efficiencies of ATP synthesis can be prepared. ATP/O quotients of up to 3 have been reported for the oxidation of NADH, and within the range 0.5 to 0.9 for the oxidation of nitrite and ascorbate-TMPD. Furthermore, NADH oxidation is associated with significant respiratory control which can be released by FCCP, or ADP plus phosphate, but not by reagents which specifically collapse only ΔpH or $\Delta\psi$. Interestingly, both the aerobic oxidation of nitrite and the reduction of endogenous cytochrome c by nitrite are inhibited by FCCP and ADP plus phosphate, or by reagents such as valinomycin $+ K^+$ which specifically collapse $\Delta\psi$ (whilst enhancing ΔpH), but are stimulated by reagents such as NH_4Cl which specifically collapse ΔpH (whilst enhancing $\Delta\psi$). It is clear, therefore, that although the overall nitrite oxidase is able to generate a Δp and hence drive ATP synthesis, the nitrite-cytochrome c reductase segment is energy-dependent. An attractive

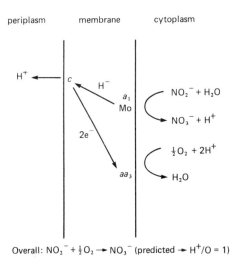

Overall: $NO_2^- + \tfrac{1}{2}O_2 \rightarrow NO_3^-$ (predicted $\rightarrow H^+/O = 1$)

Fig. 3.3 The proton-translocating nitrite oxidase in *Nitrobacter* (after Cobley, 1976). Note that at pH 8, the optimum pH for nitrite oxidation, the E_m of the nitrate/nitrite couple is $+360$ mV cf. $+327$ mV for the ferri/ferrocytochrome a_1 couple; the initial oxidation of nitrite is therefore not significantly energy-dependent. The subsequent transfer of H^- towards cytochrome c is driven by $\Delta\psi$.

Aerobic respiration in chemolithotrophs; anaerobic respiration

chemiosmotic mechanism has been put forward to explain these results whereby the initial oxidation of nitrite by cytochrome a_1/Mo is followed firstly by the outward transfer of a hydride ion towards cytochrome c, and finally by the inward transfer of $2e^-$ to molecular oxygen via cytochrome oxidase aa_3 (Fig. 3.3). H^- transfer in the first part of this modified redox loop is therefore driven by the $\Delta\psi$ component of the protonmotive force generated by the action of the entire loop. This reaction mechanism is fully supported by the redox potentials and spatial organization of the respiratory chain components. It is experimentally very difficult to check the predicted $\rightarrow H^+/O$ quotient of 1 for the oxidation of nitrite, since conditions which maximize the stoichiometry of respiration-linked proton translocation tend to inhibit nitrite oxidation. It should be noted, however, that the observed ATP/O quotient of up to 0.9 for the oxidation of nitrite by inside-out vesicles seems too high to be accommodated by the predicted $\rightarrow H^+/O$ quotient.

Reversed respiration from nitrite to NAD^+ is particularly expensive energetically since it involves sites 1 and 2 as well as nitrite-cytochrome c reductase. The reduction of one molecule of NAD^+ thus requires the total oxidation of at least 3, and probably as many as 5, molecules of nitrite. The molar growth yields of species of *Nitrobacter*, like *Nitrosomonas* species, are therefore extremely low.

The reduction of nitrate to nitrite The bacteria which catalyse the dissimilatory reduction of nitrate to nitrite are preponderantly facultative anaerobes and include *E. coli*, *K. pneumoniae*, *P. denitrificans* and several denitrifying species of *Pseudomonas*.

Several membrane-bound nitrate reductases have been extensively investigated, particularly in *E. coli* where genotypic and phenotypic manipulations are fairly easy to carry out. The *E. coli* enzyme has recently been purified and shown to be a molybdenum-containing, iron-sulphur protein which is composed of two non-identical subunits: α (MW 155000) and β (MW 63000); it is closely associated with a b-type cytochrome specific to nitrate reduction ($b_{556}^{NO_3^-}$; MW 19000) which is often regarded as the γ subunit. The overall complex has a subunit stoichiometry of $\alpha\beta\gamma$ or $\alpha\beta\gamma_2$, but probably exists as a tetramer of this *in vivo*, and contains 12 atoms each of iron and labile sulphur per atom of molybdenum. EPR studies indicate that the iron-sulphur and molybdenum centres are both involved in respiratory activity, the latter probably acting as an Mo (VI)/Mo (IV) couple. The nitrate reductases of *K. pneumoniae*, *P. denitrificans* and several other organisms appear to have approximately similar properties.

The spatial organization of the *E. coli* nitrate reductase within the coupling membrane has been extensively investigated by conventional procedures, supplemented by the use of novel non-permeant reductants such as ethylene-bipyridylium (diquat radical; DQ^+), reduced N-methylphenazonium methosulphate (PMSH), $FMNH_2$ and dibenzyl-bipyridylium (benzyl viologen radical; BV^+), all of which are readily autoxidizable and hence can only be used to investigate anaerobic systems. The results indicate that cytochrome $b_{556}^{NO_3^-}$ is located on the periplasmic surface of the membrane, whereas the α and β subunits are exposed on the cytoplasmic side, where the larger subunit catalyses both the reduction of nitrate and the consumption of $2H^+$. Furthermore, the anaerobic oxidation of DQ^+ (an electron donor), PMSH (a hydride donor) and ubiquinol-1 (a 2H donor) occurs on the periplasmic surface of the membrane and yields $\rightarrow H^+/2e^-$ quotients of approximately 0, 1 and 2 respectively in the absence of FCCP, and of approximately

Bacterial Respiration and Photosynthesis

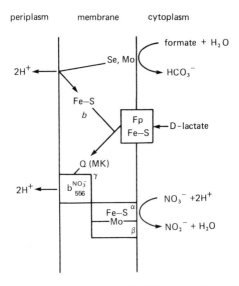

Fig. 3.4 Anaerobic respiration to nitrate in *E.coli* (after Haddock, 1977). Note that NADH dehydrogenase is also present and, like formate dehydrogenase, possibly forms a proton-translocating redox loop (see Fig. 2.4).

-2, -1 and 0 respectively in its presence. These redox reactions are not exhibited by haem-deficient mutants (hem^-), but are retained by double quinone mutants (ubi^- men^-). The overall evidence thus supports the idea that proton translocation occurs via a redox arm mechanism, involving cytochrome but not quinone, in which the proton-liberating and proton-consuming reactions occur on opposite sides of the membrane, and that the nitrate reductase complex (like cytochrome oxidase aa_3) spans the coupling membrane and catalyses transmembrane electron flow (Fig. 3.4). Similar conclusions have been drawn for the nitrate reductases of several other organisms, including *K. aerogenes*.

The physiological substrates for nitrate respiration are generally the same as those for aerobic respiration. However, during anaerobic growth *E. coli* additionally produces formate and hence synthesizes an active formate dehydrogenase which comprises part of a relatively simple formate-nitrate reductase respiratory system. Formate dehydrogenase consists of 3 subunits (α, β and γ, of MW 110 000 32 000 and 20 000 respectively) and contains selenium, molybdenum and haem in equimolar amounts, together with a large excess of iron-sulphur protein; the selenium and haem are probably associated with the α and γ subunits respectively. The molybdenum possibly constitutes the first redox centre in the system, but the role of selenium is currently unclear. Covalent labelling studies suggest that the two major subunits span the coupling membrane; the spatial arrangement of the other redox centres is unknown, but there is tenuous evidence that the formate binding site is on the cytoplasmic surface of the membrane. In contrast to the *E. coli* enzyme, the formate dehydrogenase from *Vibrio succinogenes* lacks both selenium and iron-sulphur proteins, and is located on the periplasmic side of the membrane.

Aerobic respiration in chemolithotrophs; anaerobic respiration

$\rightarrow H^+/NO_3^-$ quotients for the oxidation of NAD^+-linked substrates and/or formate by various organisms indicate the presence of two proton-translocating segments, each of which exhibits an $\rightarrow H^+$/site ratio of 2 and generates a Δp sufficient to drive ATP synthesis. These low $\rightarrow H^+/NO_3^-$ quotients ($\cong 4$) are in accord with the relatively low redox potential of the nitrate/nitrite couple compared with that of the oxygen/water couple, and have been confirmed both by the assay of ATP/O quotients in inside-out vesicles and by the measurement of molar growth yields.

The synthesis of nitrate reductase, which can account for 10–20% of the protein in the anaerobic respiratory membrane, is induced by nitrate and repressed by oxygen. Furthermore, growth under molybdenum-limited conditions, or in the presence of a molybdenum antagonist such as tungstate, leads to the synthesis of inactive formate dehydrogenase and nitrate reductase. At least nine genes appear to be required for the synthesis of formate-nitrate reductase in *E. coli* in addition to those which are responsible for the quinones and haems (Fig. 3.5). Thus Fdh$^-$ mutants do not make the main apoproteins of formate dehydrogenase (for which the *fdhA* and *fdhB* genes are responsible) whereas Chl$^-$ mutants (so-called because of their resistance to inhibition by chlorate, an analogue of nitrate) are defective in nitrate reductase or both enzymes. These latter phenotypes respectively reflect the synthesis of inactive nitrate reductase apoproteins (probably *chlI*, *chlC* and *chlG*) and the inability to insert molybdenum into the apoproteins of both enzymes (*chlD*, *chlA* and *chlB*). The role of the *chlE* gene is currently unclear. There is evidence that the structural genes for formate dehydrogenase and nitrate reductase are scattered around the chromosome and do not form discrete operons.

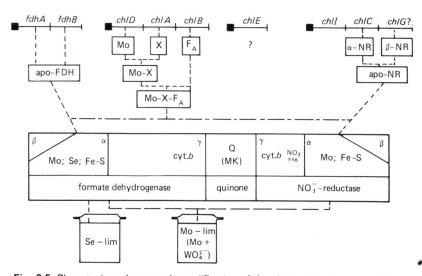

Fig. 3.5 Phenotypic and genotypic modification of the nitrate-reducing, anaerobic respiratory chain of *E. coli* (after Haddock, 1980). The manipulations which affect the synthesis of quinones and haems are not included (see Fig. 2.1).

Bacterial Respiration and Photosynthesis

The reduction of nitrite to molecular nitrogen The nitrite produced by the action of nitrate reductase is potentially toxic to bacterial cells and is exported through the cytoplasmic membrane, possibly via the combined action of two separate symports: an $H^+.NO_2^-$ symport which exports nitrite, and an $H^+.NO_3^-$ symport which imports nitrate. The overall system thus acts as an electroneutral $NO_2^-.NO_3^-$ antiport which effectively balances the inflow of substrate and the outflow of product.

The exported nitrite is reduced to nitrous oxide (almost certainly via nitric oxide) and hence to molecular nitrogen by the enzymes nitrite reductase and nitrous oxide reductase respectively, both of which are present in the periplasm or are loosely attached to the periplasmic surface of the membrane. In *P. denitrificans* and other denitrifying bacteria such as *P. aeruginosa* these dissimilatory reactions occur via anaerobic respiration, the reductases being induced by nitrate (either added exogenously or derived endogenously via the reduction of nitrate) and/or nitrous oxide. In addition, their activities are inhibited by oxygen, but the mechanism of this phenomenon is unknown.

Respiratory nitrite reductase is a specialized cytochrome cd_1 (MW 122,000) which is composed of two identical subunits, each of which binds one molecule of haems c and d_1. This soluble enzyme exhibits some cytochrome oxidase activity, and indeed resembles cytochrome oxidase aa_3 in having four one-electron redox centres. However, its affinity for (and rate of reaction with) oxygen is much lower than for nitrite, and there is little doubt that its major function *in vivo* is to reduce nitrite to nitrous oxide. Electron transfer from the Q-b region of the respiratory chain to nitrite involves two c-type cytochromes and a blue copper-containing

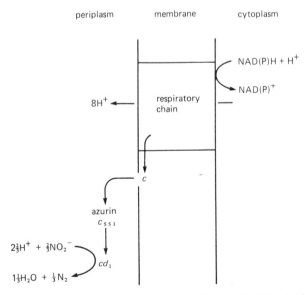

Fig. 3.6 Anaerobic respiration to nitrite in *Pc. denitrificans* (after Wood, 1978; and Meijer *et al.* 1979). The overall 3e⁻ reduction of nitrite to molecular nitrogen is normalized to a 2e⁻ transfer process.

Aerobic respiration in chemolithotrophs; anaerobic respiration

protein, *azurin* (MW ≤ 15000; $E_m \geq +230$ mV). The latter contains one atom of copper and transfers electrons either between the membrane-bound cytochrome c and the periplasmic c_{551}, or between the latter and nitrite reductase. There is some evidence that nitrous oxide reductase is similarly linked to cytochrome c, but little is known about the properties of this enzyme except that its activity is readily inhibited by acetylene (HC$\equiv$CH).

Nitrite respiration is accompanied by net electrogenic proton translocation (Fig. 3.6), and hence by ATP synthesis, despite the fact that the overall reduction of one molecule of nitrite to half a molecule of nitrogen consumes 4H$^+$ on the periplasmic side of the membrane. $\rightarrow$H$^+$/NO$_2^-$ quotients of approximately 8 and 2 have been reported for the oxidation by *P. denitrificans* of endogenous substrates (NADH and/or NADPH) and succinate respectively, and are in accord with the values that can be predicted for this 3e$^-$ transfer process, i.e.

$$\rightarrow H^+/NO_2^- = 1.5(\rightarrow H^+/O) - 4$$

These results are supported by the molar growth yields, which also indicate that each of the sequential reactions $NO_2^- \rightarrow \frac{1}{2}N_2O \rightarrow \frac{1}{2}N_2$ conserves an equivalent amount of energy per electron transferred.

In *E. coli*, *K. aerogenes* and some other facultative anaerobes, nitrite is reduced to ammonia by a soluble NADH-dependent nitrite reductase. This reaction, which is responsible for the short-circuiting of the nitrogen cycle described above, does not directly conserve energy, but by reoxidizing NADH it shifts fermentative metabolism towards the production of acetate rather than ethanol, and hence brings about a slight increase in the yield of ATP via substrate-level phosphorylation.

Nitrogen fixation Unlike nitrification and denitrification, nitrogen fixation is associated with the hydrolysis rather than the synthesis of ATP. It is catalysed by an oxygen-sensitive enzyme complex, nitrogenase, which reduces one molecule of molecular nitrogen to two molecules of ammonia at the expense of ATP and a suitable low redox potential reductant. The synthesis of nitrogenase is repressed by ammonia and oxygen, and the latter also irreversibly inhibits the activity of the purified enzyme. Nitrogen fixation is therefore an essentially anaerobic process and as such is principally confined to obligate anaerobes (e.g. *Clostridium pasteurianum* and *Chromatium vinosum*) and to facultative anaerobes growing anaerobically or at low oxygen tensions (e.g. *K. pneumoniae*, *B. polymyxa*, *R. rubrum*). The few species of bacteria which are able to fix nitrogen aerobically (e.g. *Azotobacter* species, various blue-green bacteria, and symbionts such as *Rhizobium* species) have developed sophisticated methods of protecting their nitrogenase from inactivation or repression by oxygen. For example, the genus *Azotobacter* exhibits an immediate dual response to oxygen stress in which (i) it decreases its intracellular oxygen concentration by enhancing the rate of respiration through its branched respiratory chain (respiratory protection), and (ii) it reversibly inactivates its nitrogenase, whilst at the same time stabilizing it against the deleterious effects of oxygen, by complexing it with a small [2Fe-2S] protein (conformational protection).

In obligate and facultative anaerobes, nitrogenase receives electrons from a reduced Fe-S protein, ferredoxin, generated during the oxidative decarboxylation of pyruvate to acetate (the phosphoroclastic reaction) or via photosynthetic electron transfer. In contrast, aerobic nitrogen fixation in the genus *Azotobacter* is characterized by energy-dependent electron transfer from NADH via novel redox carriers, including flavodoxin and NADH-flavodoxin reductase, and hence may be

Bacterial Respiration and Photosynthesis

regarded as a specialized form of reversed respiration:

$$12ATP + N_2 \longrightarrow 12Pi + 12ADP + 2NH_3$$
$$6H \longleftarrow 3NADH + 3H^+ / 3NAD^+$$

Nitrogenase has been purified from several species of bacteria and in each case has been shown to consist of an Fe protein and an Mo. Fe protein, both of which are oligomeric. The former contains a single [4Fe-4S] centre, whereas the latter contains 2 atoms of molybdenum and approximately 28 atoms of iron and labile sulphur that are organized into several [4Fe-4S] centres plus two [3Fe-Mo-4S] centres.

The mechanism of action of nitrogenase is extremely complex and can only be considered in the briefest outline here. A highly simplified view envisages that the Fe protein (complexed with $Mg.ADP^-$ or $Mg.ATP^{2-}$ at different stages of the reaction) transfers $2e^-$ from reduced ferredoxin or flavodoxin to the Mo.Fe protein, with the concomitant hydrolysis of 4ATP, and that the Mo.Fe protein subsequently transfers these electrons to N_2 which simultaneously binds $2H^+$. Three successive cycles of this type generate $2NH_3$, possibly via Mo-bound N_2-hydride intermediates (e.g. $Mo = N-NH_2$). It should be noted that in the absence of N_2, nitrogenase catalyses the reduction of H^+ to hydrogen, or acetylene to ethylene. The latter is a particularly useful reaction experimentally since it enables nitrogenase activity to be monitored using gas-liquid chromatography.

Considerably less is known about electron transfer from NADH to flavodoxin. The latter is a small FMN-containing flavoprotein (MW $\simeq$ 12000), the apoprotein of which closely resembles that of ferredoxin; indeed, flavodoxin is often synthesized instead of ferredoxin during growth under iron-limited conditions. Like most flavoproteins, flavodoxin can exist in the oxidized, semiquinone, and reduced (hydroquinone) states, but during electron transfer to nitrogenase it oscillates only between the latter two forms (E_m $FvdH/FvdH_2 \cong -440$ mV). There is compelling evidence that reversed electron transfer from NADH to FvdH, and hence to nitrogenase, is driven by the $\Delta \psi$ component of the protonmotive force generated by aerobic respiration. $\Delta \psi$ also appears to exert a fine control over this process, the rate of nitrogen fixation falling to zero as $\Delta \psi$ is experimentally decreased from 110 to 80 mV, with Δp being held essentially constant. This threshold phenomenon is particularly interesting since, at least in *A. vinelandii*, it can be initiated by low concentrations of ammonium ions (the uptake of NH_4^+ via a $\Delta \psi$-dependent uniport collapses $\Delta \psi$ and thus provides a mechanism via which the energetically expensive process of nitrogen fixation can be switched off immediately it is not required). Little is known about the composition of NADH-flavodoxin reductase, although an iron-sulphur protein has been implicated in this enzyme. The components of this low redox potential respiratory chain appear to undergo coordinate repression and derepression in parallel with nitrogenase.

The oxidation and reduction of sulphur compounds

The sulphur cycle The dissimilatory oxidation and reduction of sulphur compounds (sulphurification and desulphurification) occurs via consecutive electron transfer reactions which convert sulphide (oxidation state -2) to sulphate ($+6$) and back again (Fig. 3.7). The oxidative sulphurifying half of the cycle can occur

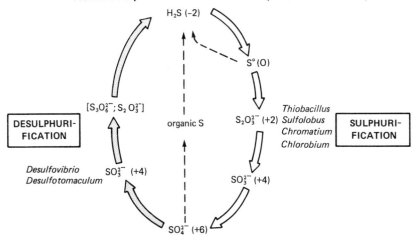

Fig. 3.7 The sulphur cycle. The dotted lines represent the reductive assimilation of sulphate and the subsequent decomposition of organic sulphur compounds; the dashed line indicates a short-circuit via which *Desulfuromonas* reduces sulphur to sulphide. Note that the phototrophic bacteria tend to accumulate elemental sulphur until all of the available sulphide has been oxidized.

either aerobically or anaerobically, whereas the reductive desulphurifying half is a completely anaerobic series of reactions. The sulphur cycle has been less extensively studied than the nitrogen cycle, mainly because of its relatively complicated chemistry and the ease with which many of its constituent compounds undergo non-enzymic transformations; in addition, some of the organisms responsible are difficult to grow in pure culture. Nevertheless, it is clear that it involves the same basic types of redox reactions as are present in the nitrogen cycle, i.e. reduced

Table 3.4 The E'_θ values of redox couples involved in sulphurification and desulphurification

Sulphurification		Desulphurification	
Redox couple	E'_θ(mV)	Redox couple	E'_θ(mV)
		AcetylCoA + CO_2/Pyr + CoASH	−490
SO_4^{2-}/SO_3^{2-}	−480	(SO_4^{2-}/SO_3^{2-}	−480)
		$2H^+/H_2$	−420
S^0/S^{2-}	−280		
		Pyruvate/Lactate	−190
		SO_3^{2-}/S^{2-}	−116
$APS^{2-}/AMP^{2-} + SO_3^{2-}$	−60	$APS^{2-}/AMP^{2-} + SO_3^{2-}$	−60
NO_3^-/NO_2^-	+420		
$\frac{1}{2}O_2/H_2O$	+820		

sulphur compounds are oxidized by a relatively high redox potential oxidant (usually oxgen but occasionally nitrate) whereas oxidized sulphur compounds are reduced by a low redox potential reductant (principally pyruvate or hydrogen). It should be noted, however, that the redox couples which comprise the sulphur cycle have lower E_0' values than those which are responsible for the nitrogen cycle (-480 to -60 mV compared with $+66$ to $+1355$ mV respectively; Table 3.4). Sulphurification reactions therefore transfer electrons over a much wider redox potential range than desulphurification reactions, and hence are responsible for liberating most of the free energy which is released during the sulphur cycle.

The oxidation of sulphide to sulphate This series of reactions is catalysed by various sulphur oxidizing bacteria including the chemolithotrophic genera *Thiobacillus* and *Sulfolobus*, and several genera of phototrophic bacteria including *Chromatium*, *Chlorobium* and *Chloroflexus* (Tables 3.1 and 4.1). However, some of these phototrophs oxidize sulphide incompletely, and all of them tend to accumulate elemental sulphur (internally in the purple bacteria, externally in the green bacteria) until all the sulphide has been oxidized. Since the oxidation of reduced sulphur compounds results in the formation of considerable quantities of acid products (e.g. SO_3^{2-}, SO_4^{2-}), particularly under aerobic conditions, some of the chemolithotrophs have developed acidophilic properties.

It appears increasingly likely that all species of *Thiobacillus*, whether growing aerobically or anaerobically with nitrate, initially cleave thiosulphate to sulphite and sulphane-sulphur [S] using the enzyme rhodanese. [S] is then oxidized to sulphite by the sulphur oxidizing enzyme, an iron-sulphur flavoprotein, and the quinone-cytochrome system to allow energy coupling at sites 2 and 3 (Fig. 3.8). Sulphite is subsequently oxidized to sulphate via two routes. The first of these involves direct oxidation by sulphite-cytochrome c reductase and the terminal cytochrome system, and results in energy coupling at site 3 under aerobic conditions (ATP/O quotients of up to 0.9 have been reported with inside-out vesicles prepared from *T. neapolitanus* and *T. novellus*, and whole cells of the former exhibit $\to H^+/O$ quotients of approximately 2). The second route is via the adenosine phosphosulphate (APS) pathway using the iron-sulphur flavoprotein APS reductase, ADP sulphurylase and adenylate kinase. During this series of reactions half a molecule of ATP is produced via substrate level phosphorylation (APS is an energy-rich phosphosulphate with a high free energy of hydrolysis; $\Delta G^{\theta'} = -88$ kJ.mole^{-1}), and two electrons enter the respiratory chain at the flavin level to generate, via oxidative phosphorylation, an additional one or two molecules of ATP depending on whether nitrate or oxygen is the terminal electron acceptor. In some species of *Thiobacillus* the last two steps in the APS pathway may be replaced by a single pyrophosphoryl transfer reaction:

$$APS^{2-} + \text{pyrophosphate}^{4-} \xrightarrow{\text{ATP sulphurylase}} ATP^{4-} + SO_4^{2-}$$

which thus doubles the ATP yield during substrate-level phosphorylation (pyrophosphate arises as an otherwise readily-hydrolysable by-product of several metabolic reactions).

Virtually nothing is known about the spatial organization of these various respiratory chains within the coupling membrane. Although the latter, unlike that of the nitrifying bacteria, is relatively free of intracytoplasmic intrusions, no reports have yet appeared which describe the magnitude or composition of any Δp which

Aerobic respiration in chemolithotrophs; anaerobic respiration

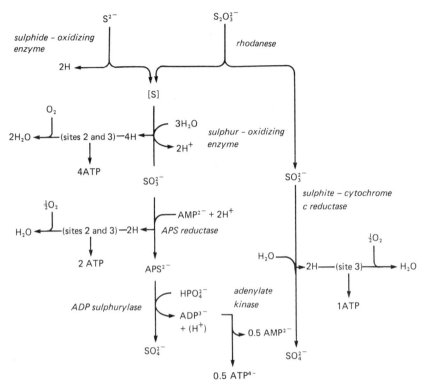

Fig. 3.8 The oxidation of thiosulphate to sulphate by various species of *Thiobacillus* (after Kelly, 1978). Note that the oxidation of one molecule of thiosulphate yields 6-9 molecules of ATP aerobically and 2-5 molecules of ATP during nitrate respiration; these values are commensurate with the measured molar growth yields.

may be generated during the oxidation of reduced sulphur compounds by chemolithotrophs. Uncoupler-sensitive reversed electron transfer between thiosulphate or sulphite and NAD^+ has been demonstrated in several species of *Thiobacillus*, and these chemoautotrophs thus exhibit predictably low molar growth yields. However, compared with the nitrifying bacteria they oxidize their reductants (particularly thiosulphate) relatively rapidly, and hence exhibit commensurately fast growth rates.

The reduction of sulphate to sulphide This transformation is catalysed by obligate anaerobes of the genera *Desulfovibrio* and *Desulfotomaculum*. These organisms generally inhabit muddy environments where they are responsible for the odoriferous accumulation of hydrogen sulphide and for the deposition of various metal sulphide ores. Lactate or pyruvate are the most commonly used sources of carbon and reducing power, and hydrogen can be used as an electron donor by some species, but since none of these organisms are autotrophic the hydrogen must be accompanied by a suitable carbon source such as acetate. The latter can also be used by a newly discovered anaerobic genus, *Desulfuromonas*, the respiratory chain of which reduces sulphur rather than sulphate.

The initial reduction of sulphate to sulphite poses a redox problem analogous to that which accompanies the oxidation of ammonia to hydroxylamine during nitrification, i.e. the E'_0 of the SO_4^{2-}/SO_3^{2-} couple is too low to allow sulphate to act as an electron acceptor. This problem is overcome by priming sulphate at the expense of ATP via a reversal of the ATP-sulphurylase reaction described above. The adenosine phosphosulphate so produced is a much more suitable oxidant since the E'_0 value of the $APS^{2-}/AMP^{2-} + SO_3^{2-}$ couple is approximately 420 mV higher than that of the SO_4^{2-}/SO_3^{2-} couple. Since the second product of this reaction, pyrophosphate, is subsequently hydrolysed to yield two molecules of orthophosphate, the overall priming reaction consumes two ATP equivalents. The reduction of APS^{2-} to sulphite is catalysed by APS reductase, an oligomeric iron-sulphur flavoprotein (MW 220 000) which contains one molecule of FAD and 12 atoms each of iron and labile sulphur.

The further reduction of sulphite is catalysed by various genera-specific bi-sulphite reductases (so called because at their pH optima of approximately 6, SO_3^{2-} exists as HSO_3^-). Two major bisulphite reductases are known, desulfoviridin and pigment P_{582}, the former being present in almost all species of *Desulfovibrio* and the latter in *Desulfotomaculum*; a third bisulphite reductase, desulforubidin, is confined to the Norway strain of *Dv. desulfuricans*. All three contain up to 14 atoms each of iron and labile sulphur, plus two molecules of sirohaem (a modified protohaem) or a close relative. In principle the reduction of sulphite to sulphide can occur either via a concerted six-electron reaction catalyzed by bisulphite reductase:

$$SO_3^{2-} + 6e^- + 6H^+ \longrightarrow S^{2-} + 3H_2O$$

or via three successive two-electron transfers which form trithionate ($S_3O_6^{2-}$) and thiosulphate ($S_2O_3^{2-}$) as intermediates, and which require the additional presence of trithionate reductase and thiosulphate reductase:

$$3SO_3^{2-} \xrightarrow[6H^+]{2e^-, \; 3H_2O} S_3O_6^{2-} \xrightarrow[SO_3^{2-}]{2e^-} S_2O_3^{2-} \xrightarrow[SO_3^{2-}]{2e^-} S^{2-}$$

The various bisulphite reductases have the redox capacity for storing and transferring six electrons, and will indeed reduce sulphite completely to sulphide, albeit releasing variable amounts of trithionate and thiosulphate according to the exact assay conditions. On the other hand, a highly active thiosulphate reductase has been purified from *Desulfovibrio*, and crude cell extracts will catalyse the reduction of trithionate. The mechanism of sulphide formation is therefore unproven, although the overall evidence probably favours a $6e^-$ transfer *in vivo*.

Respiration from the appropriate primary dehydrogenases (e.g. pyruvate-ferredoxin reductase, lactate dehydrogenase and hydrogenase) to the APS and bisulphite reductases is catalysed by a specialized low redox potential respiratory chain which contains ferredoxin, flavodoxin, menaquinone, cytochrome b and, in the genus *Desulfovibrio* but not in the genus *Desulfotomaculum*, cytochrome c_3. Ferredoxin and flavodoxin, which are functionally interchangeable, accept reducing equivalents from pyruvate and later on in the respiratory chain transfer them to the APS and bisulphite reductases. This dual site of action reflects the abilities of

Aerobic respiration in chemolithotrophs; anaerobic respiration

both of these redox carriers to operate at two different redox potentials; ferredoxin can exist in both a trimeric (FdI; $E_m = -440$ mV) and a tetrameric (FdII; $E_m = -140$ mV) form, whereas flavodoxin can oscillate between its oxidized and semiquinone forms (E_m Fvd/FvdH $= -150$ mV), and between its semiquinone and hydroquinone forms (E_m FvdH/FvdH$_2$ $= -440$ mV). Cytochrome c_3 is a novel haemoprotein (MW 13 000) that usually contains four haems c (the haems each carry one electron, but exhibit different E_m values within the range -324 to -284 mV); the major function of c_3 is probably to accept electrons from hydrogenase. The sulphate-reducing bacteria also contain a monohaem cytochrome c_{553} ($E_m = +50$ mV), rubredoxin (a one-iron protein which lacks labile sulphur; $E_m = -60$ mV) and a molybdenum-containing iron-sulphur protein (several redox centres; $E_m \leq +260$ mV), but the functions of these components during sulphate respiration are currently unknown.

A major characteristic of sulphate respiration is the soluble nature of many of its constituent redox carriers. Thus, hydrogenase and cytochrome c_3 are positioned in the periplasm, and ferredoxin, flavodoxin, the various reductases and the pyruvate phosphoroclastic system are cytoplasmic; only lactate dehydrogenase, menaquinone and cytochromes b and c are embedded in the coupling membrane.

Little quantitative information is available on ATP synthesis in cell-free extracts of members of the genus *Desulfovibrio* and *Desulfotomaculum*, and no studies of respiration-linked proton translocation have been reported. However, measurements of molar growth yields indicate that these organisms can catalyse both oxidative and substrate-level phosphorylation, the former being associated with sulphate respiration and the latter with the oxidation of pyruvate to acetate (e.g. during growth on lactate or pyruvate). The results suggest that the overall reduction of sulphate to sulphide at the expense of 4H$_2$ generates 1 ATP net (i.e. 3 ATP via oxidative

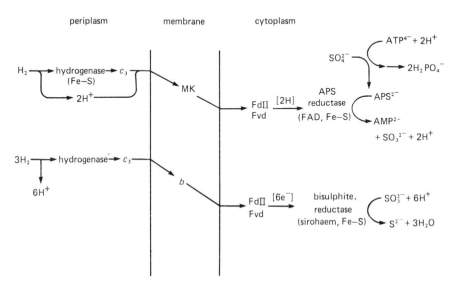

Fig. 3.9 Anaerobic respiration from H$_2$ to APS and (bi) sulphite in *Desulfovibrio* (after Wood, 1978; Thauer and Badziong, 1980).

phosphorylation minus 2 ATP equivalents used for APS formation). Energy coupling appears to be restricted to the level of site 1, and the spatial organization of the redox carriers is consistent with the presence of two types of chemiosmotic redox arm, one carrying 2H and the other only electrons (Fig. 3.9). The entry of sulphate and the exit of sulphide probably occur via separate $2H^+ . SO_4^{2-}$ symports, such that the overall system acts as an electroneutral $SO_4^{2-} . S^{2-}$ antiport which effectively balances the inflow of substrate and the outflow of product (as is probably also the case with NO_3^- and NO_2^- in denitrifying bacteria).

The oxidation of hydrogen

The respiratory chain-linked oxidation of hydrogen is an extremely widespread phenomenon. Aerobically it is characteristic of facultative chemolithotroph such as *A. eutrophus*, *P. denitrificans* and *P. saccharophila* (Table 3.1) as well as the facultative phototroph *R. capsulata* and the nitrogen-fixing chemoheterotroph *A. vinelandii* (in the latter case representing a method of recycling reducing power wastefully liberated by nitrogenase). Anaerobically it is associated with fumarate respiration in a wide variety of facultative and obligate anaerobes, with nitrate respiration in *P. denitrificans*, with the dissimilatory reduction of sulphate and carbon dioxide by various obligate anaerobes, and with photosynthetic electron transfer in most phototrophs (Table 3.2; see also Table 4.1). The reaction is catalysed by an enzyme, hydrogenase, which is either NAD^+-dependent or NAD^+-independent. In the former case it is usually an oligomeric iron-sulphur flavoprotein (typically of $MW \cong 200\,000$ and containing FMN plus several [2Fe-2S] and [4Fe-4S] centres) that is found in the cytoplasm, whereas in the latter case it is a much simpler iron-sulphur protein (typically of $MW \leq 90\,000$ and containing only 6 atoms of iron and labile-sulphur) that is present either in the membrane or the periplasm.

There is increasing evidence from the measurement of whole cell $\rightarrow H^+/O$ quotients and molar growth yields, and of ATP/O quotients in membrane vesicles, that the membrane-bound and periplasmic hydrogenases from many organisms (but probably not *P. saccharophila*) are effectively coupled to ATP synthesis at site 1 in spite of being unable to reduce NAD^+. The key to this property lies in their spatial organization, which allows them to release $2H^+$ into the periplasm and to consume $2H^+$ from the cytoplasm. This is brought about in several organisms via protonmotive redox arms in which electron transfer across the membrane is accomplished either directly via a transmembrane hydrogenase which has its hydrogen-binding site on the external surface (as in *E. coli*) or indirectly via a periplasmic enzyme that is eventually linked to a transmembrane electron transfer system (as is probably the case in *D. desulfuricans*). In contrast, there is some evidence that the hydrogenase of *P. denitrificans* activates hydrogen on the cytoplasmic surface of the membrane and hence probably translocates protons via a redox loop or pump mechanism. It is likely, therefore, that in many of these hydrogen oxidizing bacteria the generation of NADH for carbon dioxide assimilation is effectively energy-independent since NAD^+ reduction can occur either directly via an NAD^+-linked hydrogenase or indirectly via reversed electron transfer through NADH dehydrogenase at the expense of the Δp generated by an NAD^+-independent hydrogenase.

Aerobic respiration in chemolithotrophs; anaerobic respiration

The oxidation of ferrous iron ($Fe^{2+} \longrightarrow Fe^{3+}$)

The oxidation of ferrous iron has been investigated in detail only in the acidophile *Thiobacillus ferro-oxidans*. The major problem of respiration using iron as the reductant is that the E'_0 value of the Fe^{3+}/Fe^{2+} couple ($+780\,mV$; pH independent) is too close to that of the $\frac{1}{2}O_2/H_2O$ couple to allow a significant amount of free energy to be released at near neutral pH. *T. ferro-oxidans* circumvents this problem by virtue of its ability to grow at an external pH of approximately 2, which also ensures that the Fe^{2+} remains soluble. Since the internal pH is approximately 6.5, there is a ΔpH across the coupling membrane of 4.5 units ($\equiv$ approximately $270\,mV$) which can drive ATP synthesis via the proton-translocating ATP phosphohydrolase. The function of the respiratory chain in this organism is therefore to remove these retranslocated protons from the cytoplasm and thus maintain both the internal pH and the ΔpH.

The high redox potential respiratory chain of *T. ferro-oxidans* is comprised of several cytochromes c, a specialized cytochrome oxidase a_1 (two haems a; $E_m + 420$ and $+500\,mV$) and a novel copper-containing protein, rusticyanin ($E_m + 680\,mV$) which is closely related to azurin. There is good evidence that the chain is organized vectorially, since rusticyanin and one of the cytochromes c are present in the periplasm, whereas another cytochrome c and the oxygen-binding site of cytochrome a_1 are on the periplasmic and cytoplasmic surfaces of the membrane respectively. This redox arm thus catalyses inwardly-directed transmembrane electron transfer from Fe^{2+} to oxygen, with the concomitant uptake of protons from the cytoplasm (Fig. 3.10).

Little is known about the low redox potential respiratory chain which is responsible for catalysing reversed electron transfer from Fe^{2+} to NAD^+. This latter reaction requires at least 6 Fe^{2+} to be oxidized per NAD^+ reduced, and the molar growth yields of this obligate chemolithotroph are therefore extremely low.

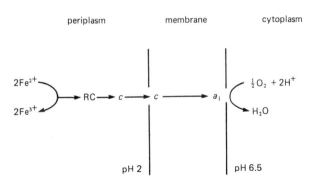

Fig. 3.10 The internal uptake of protons during the oxidation of iron by *Th. ferro-oxidans* (after Ingedew *et al.*, 1977). RC: rusticyanin.

The reduction of ferric iron ($Fe^{3+} \rightarrow Fe^{2+}$)

The use of ferric iron as a terminal electron acceptor during anaerobic respiration has recently been reported for several species of *Bacillus* and *Pseudomonas* which inhabit waterlogged soils rich in organic matter, and which can tolerate the relatively acidic conditions generated by the overall process ($[2H] + 2Fe^{3+} \rightarrow 2Fe^{2+} + 2H^+$). The high E'_0 value of the Fe^{3+}/Fe^{2+} couple would indicate energy conservation efficiencies and molar growth yields similar to those which are observed under aerobic conditions, but this has yet to be investigated experimentally. Little is known about the nature of the Fe^{3+} reductase or the organization of the respiratory chain, but for reasons of solubility it is likely that the terminal reaction occurs externally, i.e. on the outside of the membrane or in the periplasmic space. If this is so, site 3 may be inoperative (as during the reduction of nitrite).

The reduction of fumarate to succinate

The ability to catalyse fumarate respiration is fairly widespread amongst those obligately and facultatively anaerobic chemoheterotrophs which synthesize menaquinone (MK) or demethyl menaquinone (DMK), and which accumulate fumarate as an end product of anaerobic metabolism (Table 3.2). Of these organisms, *E. coli* and *V. succinogenes* have been most extensively investigated; both will use NADH, hydrogen or formate as reductant, the latter two being formed as fermentation

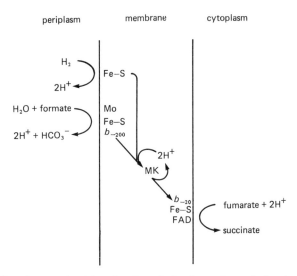

Fig. 3.11 The release and uptake of protons during fumarate respiration in *V. succinogenes* (after Kroger *et al.* 1980). Note that when NADH is the reductant this electron-transferring arm is probably preceded by an outwardly-directed 2H-transferring arm, thus forming a protonmotive redox loop (the number associated with the cytochromes *b* are their E_m values).

products. In each case, the low redox potential respiratory chain is entirely membrane-bound and consists of the appropriate primary dehydrogenases plus MK or DMK, one or more b-type cytochromes, and fumarate reductase. The latter is an iron-sulphur flavoprotein which contains two subunits (MW 79 000 and 31 000) and is closely associated with two molecules of cytochrome b (MW 25 000). It closely resembles succinate dehydrogenase, but is genetically distinct and reduces fumarate much faster than it oxidizes succinate. The failure of ubiquinone to participate in fumarate respiration probably reflects the high E_m value of the Q/QH_2 couple compared to that of the fumarate/succinate couple.

Inside-out vesicles of *E. coli* and *V. succinogenes* catalyse ATP synthesis during fumarate respiration, and whole cells exhibit respiration-linked proton translocation ($\rightarrow H^+/2e^- = 2$), ATP synthesis (ATP/O = 0.9) and solute uptake concomitant with the oxidation of endogenous substrates. These results indicate that energy coupling is confined to the level of site 1, and this has been confirmed by measurements of molar growth yields. Extensive experiments with permeant and impermeant redox couples have shown that in both organisms fumarate reductase is located on the cytoplasmic side of the coupling membrane. Thus, since the substrate-binding sites of hydrogenase and of *V. succinogenes* formate dehydrogenase both face the periplasm, it appears that energy conservation can occur via a simple, and probably chemiosmotic, redox arm mechanism (Fig. 3.11). In contrast, the substrate-binding sites of NADH dehydrogenase and of *E. coli* formate dehydrogenase are both on the cytoplasmic surface of the membrane, and there is some evidence that these two enzymes combine with the redox arm to form a protonmotive redox loop, although a proton pump mechanism cannot be entirely ruled out.

The reduction of carbon dioxide to methane

Relatively little is known about the dissimilatory reduction of carbon dioxide to methane (methanogenesis). This process is confined to a few obligately anaerobic organisms of which the genus *Methanobacterium* has been most extensively investigated. Some of these methanogens are found in the animal rumen and others occupy similar habitats to those of the sulphate-reducing bacteria, i.e. they all grow in the proximity of hydrogen-producing cellulolytic bacteria. Since hydrogen is the most frequently used reductant for methanogenesis, these organisms may be regarded as chemolithotrophs. Furthermore, since they have many properties in common with some thermoacidophiles and extreme halophiles, including the presence of unusual membrane lipids and the absence of peptidoglycan from their cell walls, all three groups have been classified as *Archaebacteria*.

The reduction of carbon dioxide to methane is an $8e^-$ transfer process which is characterized by the release of a substantial amount of free energy ($\Delta G^{\theta'} = -131$ kJ. mole^{-1}). Interestingly, the E'_θ values of the CO_2/formate and formate/formaldehyde couples (-432 mV and -535 mV respectively) are significantly lower than that of the hydrogen electrode, whereas those of the formaldehyde/methanol and methanol/methane couples (-182 mV and $+169$ mV respectively) are considerably higher. Although these intermediates are probably enzyme-bound *in vivo*, there is increasing evidence that the initial stages of methanogenesis involve reversed respiration and that the terminal stages are associated with forward respiration, energy released by the latter being used to

Bacterial Respiration and Photosynthesis

drive the former (a somewhat analogous type of coupling characterizes nitrite oxidation in the genus *Nitrosomonas*).

In addition to hydrogenase, several novel redox compounds have been implicated in methanogenesis. These include factor F_{420} (an analogue of FMN that contains a modified isoalloxazine nucleus and has a lactyl-diglutamyl moiety attached to the phosphate group; $E_m F_{420}/F_{420} H_2 = -373\,mV$), coenzyme M (2-mercaptoethane sulphonate, $HS\text{-}CH_2.CH_2.SO_3^-$, which can also exist in the oxidized dithio form; $E_m (S\text{-}CoM)_2/2\,HS\text{-}CoM = -193\,mV$) and methanopterin (a pterin derivative which was originally called B_O; $E_m\,MP/MPH_2 \cong -450\,mV$) plus several other compounds which have not been extensively purified or analyzed. No quinones or cytochromes appear to be involved in H_2-dependent methanogenesis. There is some tentative evidence that hydrogenase, methanopterin and F_{420} are involved in respiration from hydrogen to $NADP^+$ and/or the various intermediates of methanogenesis, and that CoM is associated with the transfer of methyl and hydroxymethyl groups (Fig. 3.12).

Membrane vesicles prepared from *M. thermoautotrophicum* catalyse H_2-dependent ATP synthesis. No $\rightarrow H^+/2e^-$ or ATP/O quotients have been reported, but redox potential considerations suggest that, as during sulphate and fumarate respiration, energy coupling is restricted to the level of site 1; there is also evidence

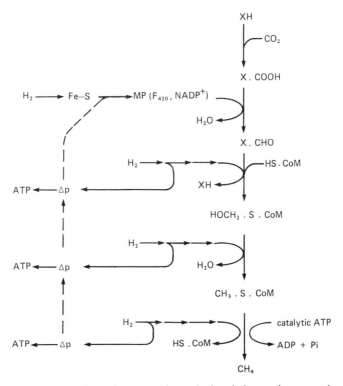

Fig. 3.12 A tentative scheme for anaerobic respiration during methanogenesis. The C_1 carrier (X) in the first two stages of the process has not yet been identified.

that hydrogen oxidation generates a Δp (negative inside). Interestingly, these vesicles contain an adenine nucleotide translocase and hence exhibit the rare property of being able to transport ATP and ADP. Indeed, there is some evidence that the methanogens may contain discrete mini-organelles (the so-called methanochondria) that are responsible for energy transduction and methanogenesis in these organisms.

The reduction of trimethlamine-N-oxide to trimethylamine

The reduction of trimethylamine-N-oxide (TMAO) to trimethylamine (TMA) occurs very widely in decaying fish where it is catalysed by various psychrophilic species of *Pseudomonas* and *Alteromonas*. The reaction has also been reported to occur in *E. coli* and other enteric bacteria, and in the facultative phototroph *R. capsulata*. Little is known about the nature of TMAO reductase or its interaction with the respiratory chain, although NADH, formate and pyruvate readily act as electron donors. The E'_θ of the TMAO/TMA couple ($+130\,\text{mV}$) is commensurate with energy coupling at least at the level of site 1, a conclusion which is supported by molar growth yield determinations.

The oxidation of carbon monoxide to carbon dioxide

This reaction is catalysed by selected species of several genera, the majority of which will also oxidize hydrogen (Table 3.1). Most of these carboxydotrophic organisms are Gram-negative aerobes, and all of them are facultative chemoautotrophs since they can use carbon dioxide as their sole source of carbon.

Little is known about carbon monoxide oxidation. However, recent investigations with *Pseudomonas carboxydovorans* have shown that the overall process, like that of nitrite oxidation, requires water and is mediated via the respiratory chain:

$$\begin{array}{c} H_2O + CO \\ CO_2 \end{array} \Bigg\rangle [2H] \longrightarrow \begin{array}{c} \tfrac{1}{2}O_2 \\ H_2O \end{array}$$

The carbon monoxide-oxidizing enzyme is a dimeric iron-sulphur protein (MW 230 000) which is not linked to $NAD(P)^+$, and is either soluble or only loosely attached to the membrane. The low E'_θ value of the CO_2/CO couple ($-540\,\text{mV}$) suggests that the enzyme interacts with the respiratory chain at a low redox potential level and that respiration ought to be accompanied by a high ATP/O quotient.

Summary

The respiratory systems of chemolithotrophs and some chemoheterotrophs are characterized by their respective abilities to oxidize inorganic reductants, and to reduce oxidants other than molecular oxygen. These reactions are catalysed by specialized respiratory enzymes, some of which contain novel redox carriers, that

Bacterial Respiration and Photosynthesis

are often inserted into conventional aerobic respiratory chains. In general, the oxidation and reduction of inorganic substrates (H_2, various nitrogen and sulphur compounds, Fe^{2+} and Fe^{3+}) occurs via metal redox centres which include iron, copper and molybdenum. The iron and molybdenum centres operate over a very wide redox potential range and are therefore found in quite diverse respiratory chains, whereas the copper centres are restricted to those chains that catalyse the oxidation and reduction of high redox potential substrates. Conversely, the anaerobic reduction of organic oxidants (fumarate, carbon dioxide) usually occurs via organic centres such as nicotinamide nucleotides, conventional or modified flavoproteins, and pterin drivatives.

Since the redox potentials of these various donors and acceptors are extremely varied, so also is the amount of free energy which is released during respiration. In the main, energy conservation during the aerobic oxidation of inorganic reductants occurs at sites 2 and/or 3 whereas during anaerobic respiration it is usually confined to sites 2 and/or 1. Energy coupling at all three sites has been confirmed only for the aerobic oxidation of hydrogen. Almost all of these respiratory chains have been shown to be protonmotive. In a few cases membrane energization probably occurs via a relatively simple redox arm mechanism which entails the release of $2H^+$ on the outside of the membrane and the consumption of $2H^+$ on the cytoplasmic side, the two protolytic reactions being linked by transmembrane electron flow (e.g. the oxidation by certain organisms of H_2 or formate concomitant with the reduction of APS, sulphite or fumarate). In other cases membrane energization is probably effected by more complex redox loop, redox cycle or proton pump mechanisms. The aerobic oxidations of nitrite and Fe^{2+} afford interesting respiratory chain variations; the former is partly energy-dependent, whereas the latter (at least in the acidophile *T. ferro-oxidans*) does not translocate protons but simply removes them from the cytoplasm as they enter via the ATP phosphohydrolase, thus maintaining the internal pH at a much higher value than that of the environment.

Many of the chemolithotrophs grow autotrophically and hence must generate NAD(P)H for the reductive assimilation of carbon dioxide. This process entails a variable amount of energy-dependent reversed electron transfer depending on the redox potential of the inorganic reductant, and is frequently at least partly responsible for the low molar growth yields that are characteristic of many of these organisms. Nitrogen fixation is a specialized form of reversed respiration, in this case from NADH to N_2 via an extremely low redox potential system which contains several novel redox carriers.

References

ADAMS, M. W. W., MORTENSON, L. E. and CHEN, J. S. (1980). Hydrogenase. *Biochemica et Biophysica Acta* 594: 105–176.

ALEEM, M. I. H. (1977). Coupling of energy with electron transfer reactions in chemolithotrophic bacteria. In: *Microbial Energetics* pp. 351–81. Edited by B. A. Haddock and W. A. Hamilton. Society for General Microbiology Symposium 27. Cambridge University Press, Cambridge.

JONES, R. W., HADDOCK, B. A. and GARLAND, P. B. (1978). Vectorial organisation of proton translocating oxido-reductions in *Escherichia coli*. In: *The Proton and Calcium Pump* pp. 71–80. Edited by G. F. Azzone. Elsevier/North Holland, Amsterdam.

KELLY D. P. (1978). Bioenergetics of chemolithotrophic bacteria. In: *Companion to*

Aerobic respiration in chemolithotrophs; anaerobic respiration

Microbiology pp. 363–86. Edited by A. T. Bull and P. M. Meadow. Longman, London and New York.

KNOWLES, C. J. (1980). (Ed) *Diversity* of *bacterial respiratory systems*. Vol 2. CRC Press, Boca Raton, Florida.

KRÖGER, A. (1978). Fumarate as terminal acceptor of phosphorylative electron transport. *Biochimica et biophysica Acta* 505: 129–45.

LEGALL, J., DerVARTANIAN, D. V. and PECK, H. D. (1979). Flavoproteins, iron proteins and haemoproteins as electron transfer components of the sulphate reducing bacteria. *Current Topics in Bioenergetics* 9: 237–65.

MORTENSON, L. E. and THORNELEY, R. N. F. (1979). Structure and function of nitrogenase. *Annual Review of Biochemistry* 48: 387–418.

ROBSON, R. L. and POSTGATE, J. R. (1980). Oxygen and hydrogen in biological nitrogen fixation. *Annual Review of Microbiology* 34: 183–207.

WOLFE, R. S. (1979). Methanogens: a surprising microbial group. *Antonie van Leeuwenhoek* 45: 353–64.

WOOD, P. M. (1978). A chemiosmotic model for sulphate respiration. *FEBS Letters* 95: 12–18.

4 Photosynthesis

Three types of bacterial photosynthesis are known:
 (i) bacteriochlorophyll-dependent (as carried out by the purple and green bacteria),
 (ii) chlorophyll-dependent (cyanobacteria or blue-green bacteria; originally called blue-green algae) and
 (iii) bacteriorhodopsin-dependent (halobacteria). Only in the blue-green bacteria is photosynthesis oxygenic (Table 4.1).

Bacteriochlorophyll-dependent photosynthesis

Photopigments and redox carriers The four families of purple and green bacteria contain light-dependent cyclic electron transfer systems that are comprised of similar basic types of photopigments and redox carriers: bacteriochlorophylls (BChl), bacteriopheophytins (BPh), carotenoids, quinones, iron-sulphur proteins and cytochromes.

Bacteriochlorophyll bears a striking resemblance to haem, except that iron is replaced by magnesium, and pyrrole ring III is fused with the adjacent methene bridge to form an additional pentanone ring (Fig. 4.1). Five types of bacteriochlorophyll are known (BChl a, b, c, d and e) which differ from each other in the nature of their peripheral groups and hence in their absorbance properties (λ_{max} 715 to 1035 nm). Each bacteriochlorophyll has a corresponding bacteriopheophytin in which the central Mg^{2+} is replaced by $2H^+$. A wide variety of carotenoids is also known (λ_{max} 400 to 500 nm) including bicyclic, monocyclic and open-chain types (e.g. β-carotene, γ-carotene and lycopene respectively), and more complex types that contain hydroxy or methoxy groups (e.g. zeaxanthin and spirilloxanthin) or at least one aromatic ring (e.g. chlorobactene).

The bacteriochlorophylls, bacteriopheophytins and carotenoids are organized within the intracytoplasmic membrane into photosynthetic units, each of which is comprised of a reaction centre complex plus multiple copies of one or two types of light-harvesting complex. The latter contain bulk amounts of various types of bacteriochlorophylls and carotenoids, and act as antennae for the absorption of radiant energy which is subsequently transferred to the reaction centre. The efficiency of energy transfer from the carotenoids to the reaction centre is much lower than from the bulk bacteriochlorophylls (20 to 90% compared with > 90%), and is thus compatible with the idea that the primary function of the carotenoids is to protect the cell from the potentially deleterious interaction of light and oxygen. The reaction centre contains bacteriochlorophylls and bacteriopheophytins in a specialized environment, possibly in association with other redox carriers. The bulk bacteriochlorophyll: reaction centre bacteriochlorophyll ratio in the photosynthetic unit ranges, according to the exact conditions of growth, from 40 to 80 in the purple bacteria and from 1000 to 1600 in the green bacteria.

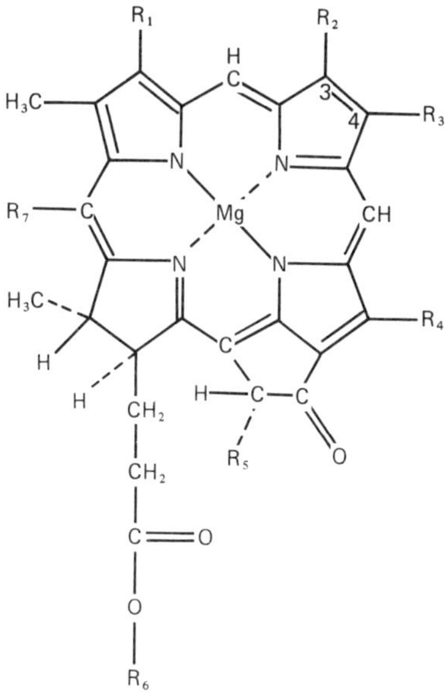

Photosynthesis

BChl	R_1	R_2	R_3	R_4	R_5	R_6	R_7	max(nm)
a	Ac	Me	Et	Me	CMe	Ph or Gg		850–910
b	Ac	Me	=CH.CH$_3$	Me	CMe	Ph		1020–1035
c	CHOH.CH$_3$	Me	Et or Pr	Et		Fa	Me	745–760
d	CHOH.CH$_3$	Me	Et or Pr	Et		Fa		725–745
e	CHOH.CH$_3$	Fo	Et or Pr	Et		Fa	Me	715–725

Fig. 4.1 Bacteriochlorophyll. Note that the double-bond between C-3 and C-4 is absent in BChl *a* and *b* Abbreviations: Ac, acetyl; CMe, carboxymethyl; Et, ethyl; Fa, farnesyl; Fo, formyl; Gg, geranylgeraniol; Me, methyl; Ph, phytyl and Pr, propyl. λmax refers to the position of the long wavelength absorption *in vivo;* in addition, all bacteriochlorophylls exhibit minor bands at 800 to 810 nm, together with major bands in the 350 to 550 nm region which overlap with carotenoid absorption. Solvent-extracted bacteriochlorophylls exhibit significantly shorter λ max values due to the absence of BChl-protein interactions.

Table 4.1 Bacteria which are capable of obtaining energy from light (phototrophs)

Family	Genus	Photopigments	Electron donor	Notes
Rhodospirillaceae (formerly *Athiorhodaceae*)	*Rhodospirillum*	BChl a; lycopene and spirilloxanthin	Organic acids	*Purple non-sulphur bacteria.* Facultatively phototrophic and photoautotrophic (*Rhodospirillum* is obligately photoheterotrophic). Some species fix nitrogen.
	Rhodopseudomonas and two other genera	BChl a or b; lycopene and spirilloxanthin	Organic acids, H_2	
Chromatiaceae (formerly *Thiorhodaceae*)	*Chromatium* and nine other genera including *Ectothiorhodospira*	BChl a; lycopene and spirilloxanthin	Organic acids, H_2, S^{2-}	*Purple sulphur bacteria.* Obligately phototrophic; facultatively photoautotrophic. Elemental sulphur accumulates within the cell (except for *Ectothiorhodospira* where deposition is external).
Chlorobiaceae (formerly *Chlorobacteriaceae*)	*Chlorobium* and three other genera	BChl a and c, d or e; Chlorobactene	Organic acids, H_2, S^{2-}	*Green sulphur bacteria.* Obligately phototrophic; facultatively photoautotrophic. Elemental sulphur accumulates and is deposited externally.
Chloroflexaceae	*Chloroflexus*	BChl a and c; β- and γ-carotene	Organic acids, S^{2-}	Facultatively photoheterotrophic. Thermophilic; some species are also alkaliphilic.
Cyanobacteriaceae kingdom (5 major subgroups)	*Anabaena* *Anacystis* *Chlorogloea* *Oscillatoria* *Nostoc* *Phormidium* and 16 other genera	Chl a; phycocyanin and/or phycoerythrin, allophycocyanin; β-carotene.	H_2O; occasionally S^{2-}	*Blue-green bacteria* (originally called blue-green algae). Obligately or facultatively phototrophic; obligately photoautotrophic. Some species fix nitrogen. Unicellar, undifferentiated filamentous and differentiated filamentous species are known.

| Halobacteriaceae | Halobacterium | Bacteriorhodopsin | None | Halobacteria. phototrophic; (carbon and reducing power obtained from amino acids). Halophilic. | Halobacteria. Facultatively heterotrophic |

Table 4.2 The simplified cyclic electron transfer system compositions of selected species of purple and green bacteria

Organism	Major light harvesting bacteriochlorophylls	Reaction centre and electron transfer components					
Rhodospirillum rubrum	BChl a (B870)	P870	BPh	QFe	Q		b c
Rhodopseudomonas capsulata	BChl a (B800, B850, B870)	P870	BPh	QFe	Q		b c
Rhodopseudomonas sphaeroides	BChl a (B800, B850, B870)	P870	BPh	QFe	Q		b c
Rhodopseudomonas viridis	BChl b (B1015)	P985	BPh	QFe	Q		b c
Chromatium vinosum	BChl a (B800, B820, B850, B890)	P870	BPh	MKFe	Q		b c
Chlorobium thiosulphatophilum	BChl c (B750), BChl a (B870)	P840	BPh			Fe-S CQ MK	b c

Abbreviations: BChl, bacteriochlorophyll; B, bulk BChl and P, reaction centre BChl (numbers represent approximate wavelengths of their major absorption bands); BPh, bacteriopheophytin; QFe, ubiquinone-iron complex; MKFe, menaquinone-iron complex, Fe-S, iron-sulphur protein; CQ, *chlorobium* quinone; other abbreviations as for Table 2.2. Note that more than one b- or c-type cytochrome is usually present in each organism.

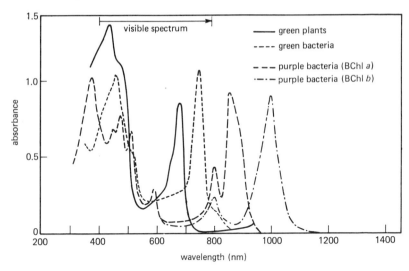

Fig. 4.2 Generalized absorption spectra of green plants and photosynthetic bacteria. Absorption below 600 nm is due to a mixture of bacteriochlorophylls and carotenoids; above this wavelength absorption is due to bacteriochlorophylls alone. The different bacteriochlorophyll contents of each group of photosynthetic organisms ensures a discrete niche for that group in the electromagnetic spectrum. Note that the purple bacteria, unlike the green plants and green bacteria, contain bacteriochlorophylls which absorb outside the visible spectrum and hence do not contribute to the colour of the organisms, which thus appear red-brown rather than green.

The cyclic electron transfer systems of the purple and green bacteria show considerable variations in their photopigment and redox carrier patterns (Table 4.2). The type of bulk bacteriochlorophyll present (BChl $a - e$) is characteristic of a particular species of photosynthetic bacterium and, in association with the carotenoid content, determines both the colour of the organism and the wavelength range of solar radiation which can be used by that organism (Fig. 4.2). Each type of bulk bacteriochlorophyll shows some variation in its absorption spectrum according to the degree of BChl-BChl, BChl-protein and BChl-carotenoid interaction. Thus, for example, BChl a exhibits varying intensities of short wavelength (B800), medium wavelength (B820 or B850) and long wavelength (B870 or B890) bands in different organisms.

Light-harvesting complexes and reaction centre complexes have recently been extracted from chromatophores of several genera of purple bacteria, most successfully from *Rhodospirillum*, *Rhodopseudomonas* and *Chromatium*. The purified light-harvesting complexes are of two types, LHI and LHII. LHI is an oligomeric protein of 12 to 15 identical subunits, plus carotenoid and long wavelength bacteriochlorophyll (B870, B890 or B1015), whereas LHII is an oligomer of up to three non-identical subunits plus carotenoid and short/medium wavelength bacteriochlorophyll (B800 + B820 and/or B800 + B850). In both complexes the BChl : carotenoid : protein ratio is approximately $\geq 2:1$ to $2:1$.

Photosynthesis

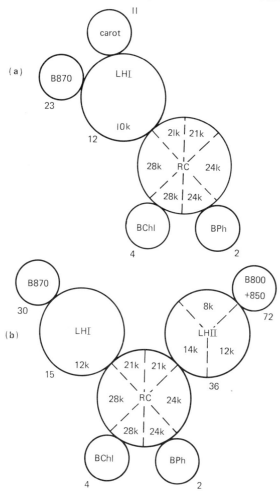

Fig. 4.3 Photosynthetic units. (**a**) *Rhodospirillum rubrum* (minimum MW 285,000); (**b**) *Rhodopseudomonas sphaeroides* and *Rhodopseudomonas capsulata* (minimum MW approximately 700,000). The large circles represent protein and enclose the approximate molecular weights of the constituent subunits; the small circles represent photopigments (note that no carotenoid contents are available for *Rhodopseudomonas* species). The numbers outside each type of circle represent the approximate number of copies per reaction centre complex. An additional species of oligomeric protein is probably present in LHII from *Chromatium* in order to accommodate B800 + B820. The photosynthetic apparatus of *Chlorobium* contains a BChl *a*-protein complex which acts between the light-harvesting and reaction centre complexes.

Bacterial Respiration and Photosynthesis

In contrast the purified reaction centre complexes consist of a single oligomeric protein of 3 non-identical subunits in the ratio $L_2M_2H_2$, plus four molecules of bacteriochlorophyll (two of which are photochemically active and are termed P870 or P985) and two molecules of the corresponding bacteriopheophytin. Variable amounts of quinone, iron, c-type cytochrome, carotenoid and phospholipid are also present depending on the organism examined and the exact extraction-purification procedure employed. The L and M subunits of the protein are associated with the photopigments and adjacent redox components, whereas the H subunits probably have a structural role in the organization of the complex (Fig. 4.3)

Another type of bacteriochlorophyll-protein complex is present in *Chlorobium* and other green bacteria, in which the protein is a trimer of identical subunits, each of which encloses 7 molecules of bacteriochlorophyll a. The unique function of this complex is apparently to transfer excitation energy from the light-harvesting apparatus in the chlorosomes to the reaction centre complex which, together with the electron transfer system, is located in the adjacent plasma membrane.

The photopigments in the reaction centre complex subsequently lose an excited electron to a variety of oxidized redox carriers which include a quinone-iron complex of unknown structure (QFe in the *Rhodospirillaceae*, MKFe in the *Chromatiaceae*; $E_m \leq -160$ mV) and an exceptionally low redox potential iron-sulphur protein (green bacteria; $E_m = -540$ mV).

All of these cyclic electron transfer systems also contain a free quinone-ubiquinone in the *Rhodospirillaceae*, menaquinone plus ubiquinone in the *Chromatiaceae*, and menaquinone plus chlorobium quinone (CQ) in the green bacteria: chlorobium quinone ($E_m + 40$ mV) is similar to menaquinone but lacks the first methylene group in the isoprenoid side chain. A high redox potential iron-sulphur protein, one or more b-type cytochromes and a cytochrome c (E_m in the range $+145$ to $+350$ mV) are present. The quinone-cytochrome system bears a striking resemblance to that present in some respiratory chains and, together with the reaction centre complex, is responsible for the generation of the energized protons that drive ATP synthesis.

The purple and green bacteria also catalyse non-cyclic electron transfer to reduce NAD^+ at the expense of various physiological substrates (e.g. sulphide, thiosulphate, organic acids, H_2) of widely differing redox potential ($E'_0 2H^+/H_2 - 420$ mV cf. fumarate/succinate $+30$ mV). Thus, as in the chemolithotrophs, reduction of NAD^+ may or may not require energy depending on the relative redox potentials of the reducing couple and the $NAD^+/NADH$ couple ($E'_0 = -320$ mV). According to the precise pathways involved, electron transfer will be catalysed by some or all of the redox carriers outlined above, together with additional membrane-bound dehydrogenases (e.g. succinate dehydrogenase, NADH dehydrogenase) and reductases (e.g. sulphide-cytochrome c reductase, thiosulphate-cytochrome c reductase, ferredoxin-NAD^+ reductase and hydrogenase) (see Fig. 1.2); an energy-linked transhydrogenase is also present.

Phenotypic modifications Changes in the growth environment of the purple and green bacteria, particularly with respect to the availability of light or oxygen, can exert profound effects both on the composition of the photosynthetic unit (e.g. the LHII:LHI ratio) and on its concentration within the cell (e.g. the area of intracytoplasmic membrane per cell and the number of photosynthetic units per area of membrane). Thus, during anaerobic growth at low light intensities, the

Photosynthesis

Rhodospirillaceae exhibit a large area of membrane which contains very high concentrations of light-harvesting bacteriochlorophylls and carotenoids, whereas at high light intensities these contents are very much lower. The overall effect of these changes is thus to vary the cross-sectional area of the antennae apparatus so as to maximize photon capture at the ambient light intensity. These changes are also seen during growth at low and high oxygen tensions respectively, the effect of oxygen being dominant to that of light. In those species that lack LHII (e.g. *R. rubrum*) these two modulators affect the concentration of the photosynthetic unit but not its composition, since neither modulator can alter the LHI:RC ratio. In many species of *Rhodospirillaceae*, repression of the photosynthetic appartus by oxygen is accompanied by an increase in cytochrome oxidase content and hence in respiratory capacity, thus allowing these facultative phototrophs to conserve energy via oxidative phosphorylation. Light is not required for the synthesis of photopigments, since phototrophs capable of fermentation remain pigmented during anaerobic growth in the dark.

Although much less is known about the detailed effects of light and oxygen on the other purple and green bacteria, these organisms appear to respond in a similar manner to the *Rhodospirillaceae*. However, since most of them lack a competent respiratory system, their growth is inhibited at high oxygen tensions; with the exception of *Chloroflexus* they are therefore classed as obligate phototrophs.

The exact mechanism by which synthesis of the photosynthetic apparatus is regulated by light or oxygen has yet to be determined.

Genotypic modification Genetic manipulation of electron transfer in purple and green bacteria has been almost completely confined to the *Rhodospirillaceae*. These facultative phototrophs readily allow the selection and screening of putative photosynthetic electron transfer mutants since the latter will usually grow aerobically in the dark, but not anaerobically in the light.

Most photosynthesis-negative (Pho^-) mutants which have so far been isolated show defects in their light-harvesting and energy-trapping apparatus. Thus, a large number of mutants with altered carotenoid contents are known, including the classical 'blue-green', 'green', 'yellow' and 'brown' mutants (such mutants are extremely useful experimentally because they allow cytochromes to be monitored spectroscopically without the usual interference from carotenoid absorption). A similarly impressive array of mutants defective in bacteriochlorophyll synthesis has also been isolated, some of which are deficient in bulk bacteriochlorophyll and others in P870. Several 'albino' mutants which lack both carotenoid and bacteriochlorophyll are also known, together with mutants which are derepressed for bacteriochlorophyll synthesis and hence are unaffected by the ambient oxygen tension. The ability of some Pho^- mutants only to grow anaerobically in the dark (i.e. fermentatively, conserving energy via substrate-level phosphorylation) suggests that one or more redox carriers are common to both photosynthesis and respiration, e.g. Q, *b, c* and a high redox potential Fe-S protein.

Gene mapping in photosynthetic bacteria has been severely restricted by the very limited capacity of these organisms for conjugation (mating), and by their apparent resistance to transformation (artificial transfer of DNA). However, the problem has been partially alleviated by the recent discovery of a gene transfer agent specific to *R. capsulata*. This agent has the morphological properties of a very small bacteriophage and is able to effect the non-lysogenic stable transduction of recipient cells, albeit to a maximum of about five genes. It has recently been used (i) to

Bacterial Respiration and Photosynthesis

confirm the presence of a branched respiratory system in *R. capsulata*, (ii) to restore the ability of Pho$^-$ mutants of this organism to synthesize light-harvesting and reaction centre bacteriochlorophylls and proteins, and (iii) to construct a gene map for photopigment synthesis. Mutations reflecting the presence of seven genes, five concerned with carotenoid synthesis (*crt A* to *crt E*) and two with bacteriochlorophyll synthesis (*bch A, bch B*), have been mapped and appear to form a tight cluster which may be under coordinate transcriptional control; in addition, the *crt* genes responsible for the various colour mutations have been delineated. It is clear that this system has considerable potential with respect to fine-structure genetics in *R. capsulata*, and should prove to be a powerful tool for analysing the mechanism and regulation of photosynthesis, respiration and nitrogen fixation in this organism. However, the size-and species-limitations of this vector emphasize the need to develop alternative methods of gene mapping in photosynthetic bacteria.

Cyclic electron transfer During the last few years considerable progress has been made towards understanding the primary events in bacterial photosynthesis, mainly through the advent of sophisticated techniques for rapid-flash illumination and for analysing the resultant changes in optical and EPR spectra. The results of such experiments with whole cells, chromatophores or purified reaction centres of selected purple bacteria, most extensively *R. sphaeroides* and *R. viridis*, indicate that illumination causes a succession of temperature-independent photochemical reactions which lead to redox changes in the reaction centre photopigments and their associated electron transfer components (Fig. 4.4).

The absorption of a single photon by the carotenoids and bacteriochlorophylls of the light harvesting complexes eventually causes singlet excitation (the raising of an electron to a higher energy level) of the longest wavelength constituent (e.g. B870). This energy is usually transferred to the reaction centre, but if the latter is inactive for some reason the energy is re-radiated as fluorescence. Under normal conditions the energy is distributed in the reaction centre between two of the four bacteriochlorophyll molecules, termed the 'special pair' ($BChl_2$ or P870), which enter the singlet excited state P870* and subsequently very rapidly transfer an electron to a single bacteriopheophytin molecule. This initial photochemical reaction between the primary donor (P870) and the primary acceptor (BPh) exhibits a quantum requirement of approximately 1, i.e. only one quantum is required for the transfer of one electron. Furthermore, since the difference in redox potential between the BPh/BPh$^-$ couple ($E_m \leq -550\,mV$) and the P870$^+$/P870 couple ($E_m \geq +450\,mV$) is approximately 1V, 70% of the energy of the incident photon (1.42 eV at 870 nm) is trapped by the system.

The electron then passes to the secondary acceptor (the ubiquinone moiety of the QFe complex; $E_m \leq -160\,mV$), after which the P870$^+$ is reduced rather slowly at the expense of the secondary donor (one of two identical *c*-type cytochromes, usually c_2; $E_m = +300\,mV$). The exergonic nature of these two latter reactions ensures that the overall photochemical process proceeds to completion and that any back reactions are kept to a minimum. The transfer of an electron from Q$^-$Fe to oxidized cytochrome c_2 occurs via the tertiary acceptor (probably a second molecule of QFe which, unlike the first, can undergo complete reduction to QH_2Fe), the bulk quinone pool and the cytochrome system, and thus causes the regeneration of the reaction centre complex in its original redox state. A similar sequence of reactions probably occurs in *R. rubrum* and *Chromatium vinosum*, except that in the latter organism MKFe replaces QFe as the secondary acceptor,

Photosynthesis

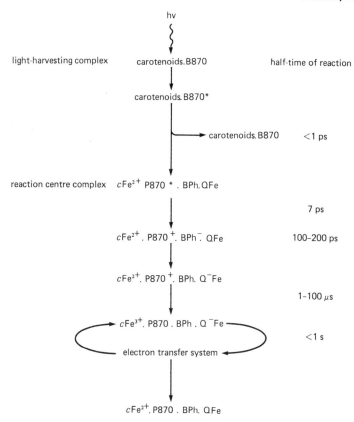

Fig. 4.4 The early stages of photosynthesis in *Rhodospirillaceae* Abbreviations: *, excited singlet state; other abbreviations as for Table 4.2. The role of the iron moiety in QFe is not known.

and cytochrome c_{555} replaces c_2 as the secondary donor. The roles of the second bacteriopheophytin molecule, the remaining two bacteriochlorophyll molecules and the iron moiety of the quinone-iron complex(es) have yet to be determined.

Relatively little is known about the early stages of photosynthesis in green bacteria. It is clear, however, that light energy absorbed by the antennae photopigments (principally BChl c and carotenoids, sometimes accompanied by BChl d and e) is subsequently transferred via bacteriochlorophyll a (B870) to the reaction centre bacteriochlorophyll (P840). This is followed by a charge separation process, the details of which have not yet been determined, that leads to the reduction of the secondary acceptor (Fe-S; $E_m = -540$ mV) and the oxidation of the primary donor (P840$^+$/P840; $E_m = +250$ mV). The primary acceptor has not yet been identified, indeed one may not be present, and the quantum requirement of the initial photochemical reaction is not known. P840$^+$ is subsequently reduced at the expense of the secondary donor (a c-type cytochrome, usually c_{555};

Bacterial Respiration and Photosynthesis

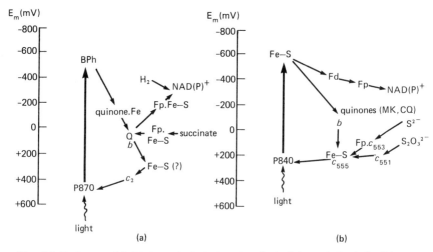

Fig. 4.5 Pathways of light-dependent electron transfer in (a) purple bacteria. (b) green bacteria. Note that cytochrome c_2 is replaced in the *Chromatiaceae* by cytochrome c_{555}; little is known about the mechanism of NAD^+ photoreduction by reduced sulphur compounds in this family.

$E_m = +145\,mV$) and the cycle is completed by electron transfer from the reduced iron-sulphur protein to the oxidized cytochrome c via the quinone pool and the cytochrome system. It is interesting to note that the entire light-dependent cyclic electron transfer system of the green bacteria operates at a significantly lower redox potential than that of the purple bacteria (Fig. 4.5). There is considerable evidence that in both groups of organisms electron transfer between the reduced secondary acceptor and the oxidized secondary donor occurs over a redox potential span of approximately 650 mV and, at least *in vitro*, is associated with electrogenic H^+ translocation and hence the generation of Δp.

Photoreduction of NAD^+ All species of purple and green bacteria require NAD(P)H for biosynthetic purposes. During photoautotrophic growth on hydrogen, this requirement is satisfied by the light-independent reaction. In contrast, during photoheterotrophic growth, the reduction of NAD^+ by various relatively high redox potential substrates is usually light-dependent, the exact pathway of electron transfer being determined by the redox potential of the primary acceptor in the photochemical process relative to that of $NAD^+/NADH$. The quinone-iron complex in the *Rhodospirillaceae* is insufficiently electronegative to reduce NAD^+ directly, and these organisms therefore use cyclic electron transfer to generate Δp which in turn drives reversed electron transfer from succinate to NAD^+ (Fig. 4.5a). This view is supported experimentally by the observations that chromatophores catalyse (i) NAD^+ photoreduction by succinate that is inhibited by uncoupling agents, ADP plus inorganic phosphate, or inhibitors of cyclic electron transfer such as antimycin A, but not by inhibitors of the ATP phosphohydrolase, and (ii) a dark reduction at the expense of ATP or pyrophosphate that is inhibited by uncoupling agents and energy transduction inhibitors, but not by inhibitors of

Photosynthesis

cyclic electron transfer. The reduction of NAD^+ by succinate is a completely membrane-bound process which probably involves succinate dehydrogenase, the Q-b region of the cyclic electron transfer system and NADH dehydrogenase (both the light- and ATP/pyrophosphate-dependent reactions are inhibited by rotenone). The reduction of $NADP^+$ by NADH is catalysed by an energy-dependent transhydrogenase, which contains no redox carriers.

The iron-sulphur protein that acts as the primary acceptor in green bacteria has a much lower E_m value than the quinone-iron complex and is thus able to reduce NAD^+ directly, without the mediation of Δp. Reduction of NAD^+ by sulphide or thiosulphate in cell-free extracts of green bacteria therefore occurs only in the light, and is not inhibited by uncoupling agents. Electron transfer from sulphide and thiosulphate to cytochrome c_{555} is mediated by two soluble reductases which contain low redox potential c-type cytochromes ($E_m \cong +180$ mV), i.e. flavocytochrome c_{553} (which also contains FMN) and cytochrome c_{551}; similarly, electron transfer from the primary acceptor to NAD^+ is dependent on a soluble, flavin-containing ferredoxin-$NADP^+$ reductase (Fig. 4.5b).

Light-induced proton translocation and phosphorylation There is now considerable evidence, particularly from experiments with *Rhodospirillaceae*, that photosynthetic electron transfer causes electrogenic H^+ translocation across the coupling membrane. On illumination, H^+ translocation occurs until a steady state condition is attained (i.e. the rate of light-induced H^+ movement is matched by the rate of H^+ back-flow); when the light is switched off, only H^+ retranslocation is possible and the pH gradient is slowly dissipated. The direction of light-dependent H^+ translocation, like that accompanying respiration, is outwards in whole cells and right-side-out membrane vesicles, and inwards in chromatophores. Since cyclic electron transfer effects no net oxidation or reduction (i.e. no exogenous donors or acceptors are involved), it is virtually impossible to measure $\rightarrow H^+/e^-$ quotients during continuous illumination; a stoichiometry based on one-electron transfer is traditionally used to describe photosynthetic proton translocation because the photochemical process separates one charge. However, $\rightarrow H^+/e^-$ quotients of approximately 2 have recently been determined using sophisticated experimental techniques that involve the measurement of H^+ translocation by chromatophores as a result of a single or repeated very brief (20 μs) flash of light, i.e. each flash is enough only to cause one complete turnover of the photosynthetic apparatus. Each electron transferred from Q^- Fe to c_2^+ appears to cause the sequential binding by the quinone system of two protons (H_I^+ and H_{II}^+) from the external aqueous phase and their subsequent release into the internal compartment, or lumen, of the chromatophore (Fig. 4.6).

Interestingly, illumination of whole cells or chromatophores of purple bacteria also causes a reversible, temperature-independent shift towards the red end of the spectrum of the major absorption bands of the bulk carotenoids. This small and kinetically complex band-shift is limited to approximately 10 to 20% of the total carotenoid content, and is divisible into three distinct phases (I, II and III) which can possibly be correlated with three separate electron transfer steps, i.e. P870$\rightarrow$QFe, ferrocytochrome $c_2 \rightarrow$ P870$^+$, and ferrocytochrome $b \rightarrow$ ferricytochrome c_2 respectively. Phases I and II are extremely rapid ($t_\frac{1}{2} < 1$ μs) compared with H^+ binding ($\cong 150$ μs) or phase III ($\cong 1$ ms). The band-shift can also be induced in dark chromatophores by the addition of ATP or pyrophosphate, or by the artificial imposition of a membrane potential ($\Delta \psi$); in the latter case it has been shown that

Bacterial Respiration and Photosynthesis

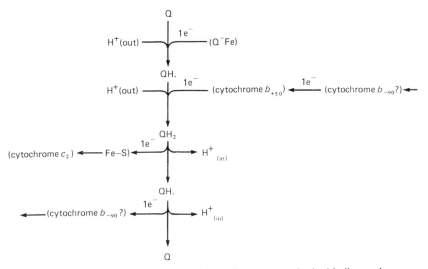

Fig. 4.6 A simplified view of the role of the quinone system in the binding and release of H^+ by chromatophores of purple bacteria. Overall, one electron is transferred from Q^- Fe to ferricytochrome c_2 (Fe^{3+}) via the Q/QH· and QH·/QH$_2$ redox couples with the inward translocation of $2H^+$. The role of cytochrome b_{-90} in the protonmotive quinone cycle is largely speculative (the numbers refer to the E_m values of the b-type cytochromes).

the direction and extent of the band shift is dictated respectively by the polarity (positive inside, → red; negative inside, → blue) and magnitude ($\Delta\lambda \propto \Delta\psi$, 1 nm per 137 mV) of the $\Delta\psi$. These observations are consistent with an electrochromic effect, i.e. a wavelength change in response to an electric field, the latter being generated either very rapidly via charge separation reactions initiated by photochemical events in the reaction centre complex (I and II) or more slowly via delocalized protonmotive forces (III). A light-dependent red-shift has also been observed in the bacteriochlorophyll absorption spectra of purple bacteria, but this phenomenon, in contrast to the carotenoid band-shift, may occur in response to ΔpH rather than $\Delta\psi$.

The effects of H^+-conductors, permeant anions and ionophorous antibiotics on light-induced proton translocation are basically similar to their effects on the respiration-linked process. Thus, both the rate of light-induced proton translocation and the extent of the resultant pH gradient are stimulated by SCN^- or valinomycin + K^+ (at the expense of $\Delta\psi$), but are diminished by FCCP, nigericin + K^+, valinomycin + nigericin + K^+, or, in chromatophores only, by NH_4^+ salts (which enhance $\Delta\psi$). There results clearly confirm that the light-induced Δp, like that generated by respiration, is composed of ΔpH and $\Delta\psi$ (internal compartment alkaline and electrically negative in whole cells and right-side-out vesicles, but acidic and electrically positive in chromatophores). Δp has recently been quantitated using standard techniques, and $\Delta\psi$ and ΔpH have also been estimated from previously calibrated band-shifts in carotenoid and bacteriochlorophyll absorption spectra respectively. The results indicate Δp values in the range 100 to 420 mV

Photosynthesis

for illuminated chromatophores or right-side-out vesicles under static-head conditions. As with respiration, the wide variations in these values of Δp perhaps reflect imperfections in the experimental procedures and/or the membrane preparations, or may indicate more localized energy coupling. Photophosphorylation is uncoupled by FCCP, valinomycin + nigericin + K^+, or NH_4^+ + a permeant anion, but not by SCN^-, valinomycin + K^+, or nigericin + K^+. These results indicate that photophosphorylation, like oxidative phosphorylation, can be driven by ΔpH or $\Delta \psi$ alone, and that uncoupling occurs only when both components of Δp are dissipated.

There appear to be a number of significant differences in the properties of chromatophores illuminated via a single-turnover flash of light compared with continuous illumination. Thus, ATP synthesis activated by a $20\,\mu s$ flash is completely inhibited by either valinomycin + K^+ or nigericin + K^+, a result which indicates that neither ΔpH nor $\Delta \psi$ alone are sufficient to drive photophosphorylation under these conditions. In addition, antimycin A causes little inhibition of ATP synthesis and only blocks the binding of H_{II}^+ ($\rightarrow H^+/e^- = 1$), whereas during continuous illumination it completely inhibits both of these processes. Some of these discrepancies between short-flash and continuous illumination have yet to be resolved, as also has the exact site (s) of action of antimycin A in the Q-b region of the redox chain.

There is increasing evidence, from a variety of experimental approaches, that the photosynthetic electron transfer systems of the *Rhodospirillaceae* (and probably also the *Chromatiaceae*) are organized asymmetrically across the intracytoplasmic membrane. Immunological studies indicate that cytochrome c_2 is loosely associated

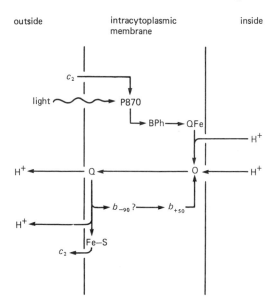

Fig. 4.7 The spatial organization of the light dependent electron transfer system of the *Rhodospirillaceae*. A total of two protons are translocated outwards during the net cyclic transfer of one electron ($\rightarrow H^+/e^- = 2$).

with the periplasmic surface of the membrane, whereas the H subunit of the reaction centre complex is embedded in the cytoplasmic face. Furthermore, detailed analyses of the effects of different photochemical charge separations (e.g. $c_2.P870^+.Q^-$ Fe cf. $c_2^+.P870.Q^-$ Fe) on the extent of the carotenoid band-shift suggest that P870 is located near the middle of the membrane and that QFe is quite close to the cytoplasmic surface. It appears likely, therefore, that the reaction centre complex spans the coupling membrane. The binding sites of the various reductases and dehydrogenases that are associated with the photoreduction of NAD^+ are on the cytoplasmic surface, or the enzymes themselves are in the cytoplasm. The evident sidedness of the cyclic electron transfer system is compatible with the gross requirements of a chemiosmotic mechanism of photosynthetic energy transduction in which the initial separation of charge across the membrane is followed by net electron and H^+ transfer in the opposite direction; the latter possibly involves a complex protonmotive quinone cycle similar to that which may be present in some respiratory chains (Fig. 4.7). Virtually nothing is known about the possible protonmotive nature or spatial organization of the photosynthetic electron transfer systems in the green bacteria.

Nitrogen fixation Many species of purple and green bacteria are able to fix nitrogen. This reaction is strictly dependent on light, the latter being required both for nitrogenase synthesis and to generate the ATP and/or the very low redox potential reductant. None of these organisms are able significantly to protect their nitrogenase from inactivation by oxygen, and hence they catalyse nitrogen fixation only under anaerobic or near-anaerobic conditions.

Chlorophyll-dependent photosynthesis

The blue-green bacteria (*Cyanobacteriaceae*) are comprised of 22 genera, and exhibit a wide range of physiological and morphological properties. Although most genera are obligately photoautotrophic, *Chlorogloea fritschii* and some species of *Nostoc* can grow photoautotrophically, photoheterotrophically or chemoheterotrophically, in the last case conserving energy via respiratory chain phosphorylation. Many genera are able to fix atmospheric nitrogen.

Cyclic electron transfer Photosynthetic energy conservation in blue-green bacteria closely resembles that in algae and green plants, i.e. it involves two photosystems (a long wavelength PSI and a shorter wavelength PSII) and is oxygenic. Cyclic electron transfer occurs via PSI, and has some superficial similarities to the same process in green bacteria. However, bacteriochlorophylls are replaced by chlorophyll *a* (Chl*a* is identical to BChl*a* except that the vinyl group at C-1 is replaced by a methoxy group), and β-carotene is the major carotenoid. There is some evidence that, at least in *C. fritschii*, these two photopigments are accompanied by small amounts of the phycobiliprotein, allophycocyanin. Solar radiation of wavelength > 660 nm is harvested by a chlorophyll *a*-protein complex (B680) and the energy is subsequently transferred to the reaction centre complex (P700), possibly via allophycocyanin; the B680:P700 ratio is approximately 45:1, and all three photopigments are situated in the thylakoid membrane. P700, like P870 probably exists as a dimer since the oxidation of this primary donor yields $(Chl)_2^+$. The identity of the primary acceptor is not known for certain, although pheophytin is currently the most favoured candidate. Iron-sulphur centres of extremely low redox potential ($E_m \leq -530$ mV) probably serve as secondary, tertiary and quaternary acceptors. Since the quantum requirement of the photochemical reaction is 1, and the E_m value of the $P700^+/P700$ complex is approximately $+450$ mV, at least 54% of the energy of the incident photon (1.82 eV at 680 nm) is trapped by PSI. Subsequent cyclic electron transfer between the reduced quaternary acceptor and the oxidized primary donor involves a pool of plastoquinone (closely related to ubiquinone, E_m $PQ/PQH_2 \leq +25$ mV), cytochromes b_{563}, b_{559} (?) and f (a c-type cytochrome), and a small copper protein, plastocyanin (E_m $PC^+/PC = +370$ mV), which acts as the secondary donor. However, this system has been little studied compared with those present in the purple and green bacteria.

Non cyclic electron transfer Non-cyclic electron transfer from water to $NADP^+$ involves photosystems I and II. The light-harvesting apparatus of PSII is composed of the phycobiliproteins phycocyanin, phycoerythrin and allophycocyanin. These consist of trimers or hexamers of two non-identical subunits ($\alpha_3\beta_3$ or $\alpha_6\beta_6$), and contain linear tetrapyrroles, bilins, as their covalently-bound chromophores. Since phycocyanin is deep blue and phycoerythrin deep red (λ_{max} 620 nm and 560 nm respectively), the relative abundance of these pigments determines both the properties of the light-harvesting antennae and the colour of the organism. These three pigments are organized into supramolecular complexes, phycobilisomes, that are comprised of a triangular core of allophycocyanin, surrounded by peripheral rods of phycocyanin and/or phycoerythrin. There is mounting evidence that these latter two pigments absorb incident light in the 550 to 660 nm range and transfer the energy to allophycocyanin (λ_{max} 650 to 670 nm); little is known about the mechanism of energy transfer between these pigments, although it is thought to occur with very high efficiency. Excited allophycocyanin subsequently transfers energy to the bulk chlorophyll (B620 + B670) and hence to the reaction centre chlorophyll (P680), both of which are located in the thylakoid membrane. PSII preparations have recently been purified from *Synechococcus* and *Phormidium*; both are enriched in short wavelength chlorophyll at the expense of the long wavelength form, and the preparation from *Phormidium* exhibits rapid, light dependent oxygen evolution. Little is known of the mechanism of this latter reaction in blue-green bacteria although, as in algae and green plants, it requires Mn^{2+} and is inhibited by dichlorophenyldimethylurea (DCMU). The high E'_θ value of the oxygen/water couple dictates that the oxidation of water requires an extremely high redox potential acceptor, and this requirement appears to be met by $P680^+$ (E_m $P680^+/P680 \geq +850$ mV), the reduced form of which becomes the primary donor for PSII.

The primary acceptor for PSII is probably plastoquinone which undergoes reduction to the semiquinone form (PQH.). Electrons are subsequently transferred via the bulk quinone pool, part of the cyclic system and PSI to the bound iron-sulphur protein, and hence via soluble ferredoxin and ferredoxin-$NADP^+$ reductase to $NADP^+$ (Fig. 4.8). Non-cyclic electron transfer from water to $NADP^+$ thus occurs over a $\Delta E'_\theta$ of approximately -1140 mV and is driven entirely by solar energy absorbed by the photosystems. This process is therefore much more akin to the photoreduction of NAD^+ by reduced sulphur compounds in the *Chlorobiaceae*, than to the Δp-driven reduction of NAD^+ by succinate in the *Rhodospirillaceae*. Indeed, *Oscillatoria limnetica* and *Anacystis nidulans* are able to reduce $NADP^+$ via

Bacterial Respiration and Photosynthesis

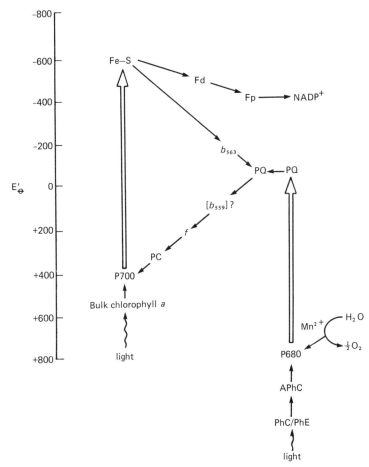

Fig. 4.8 Pathways of light-dependent electron transfer in blue-green bacteria. Abbreviations: PhC, phycocyanin; PhE, phycoerythrin; APhC, allophycocyanin; PQ, plastoquinone; PC, plastocyanin.

PSI using sulphide and thiosulphate respectively, although high sulphide concentrations inhibit the PSII of *O. limnetica* and cause photosynthesis to become anoxygenic.

Several genera of filamentous blue-green bacteria, including *Anabaena*, are capable of nitrogen-fixation when deprived of combined nitrogen, in spite of the fact that they catalyse oxygenic photosynthesis. This paradox is partly overcome by largely restricting nitrogenase to specialized cells, heterocysts, which develop from vegetative cells at regular intervals along the filament. Heterocysts catalyse only anoxygenic photosynthesis, i.e. PSII is inactive and ATP is generated solely by cyclic photophosphorylation via PSI; NADH is supplied to the heterocysts indirectly, as a glucose-containing disaccharide, from adjacent vegetative cells that

Photosynthesis

catalyse oxygenic photosynthesis. Oxygen released by the latter process is rapidly consumed via enhanced photorespiration, i.e. light-dependent oxidation (probably of NADH) via a rotenone- and cyanide-sensitive respiratory chain. There is evidence that in non-heterocystous blue-green bacteria such as *Oscillatoria*, nitrogen fixation occurs only when oxygenic photosynthesis is inactive: there is a temporal separation of the two processes; the mechanism via which this protection phenomenon operates has yet to be fully elucidated. In both groups of organisms the physiological donor to nitrogenase is probably reduced ferredoxin formed by the oxidation of NAD(P)H.

Phenotypic modifications As has been observed with the *Rhodospirillaceae*, exposure of blue-green bacteria to high light intensities causes a general decrease both in the concentration of the light-harvesting photopigments (the phycobiliproteins and chlorophylls) and in the area of thylakoid membrane. In addition, some species exhibit a differential response to light of different wavelengths, a phenomenon known as complementary chromatic adaptation, during which *de novo* synthesis of phycocyanin and phycoerythrin is stimulated by red and green light respectively such that the light-harvesting ability of the organism is maximized for each light regime. The exposure of facultatively phototrophic blue-green bacteria (e.g. *C. fritschii*) to dark conditions leads to the repression of PSII and attendant oxygen evolution, both of which are recovered on re-illumination.

Photophosphorylation Very little is known about photosynthetic energy conservation and ATP synthesis in blue-green bacteria, a situation which probably reflects the difficulty of isolating functional phycobilisome-thylakoid preparations and the resultant need to use whole cells or filaments. There is good evidence from the latter, however, that both cyclic and non-cyclic electron transfer is accompanied by ATP synthesis, and that either system alone can support maximal rates of CO_2 assimilation, nitrogen fixation and growth. ATP/e^- quotients of approximately 0.5 have been reported for several species. Indeed, there is sufficient free energy released during electron transfer between water and plastoquinone (PSII) and between plastoquinone and cytochrome *f*/plastocyanin to support ATP synthesis, and two coupling sites (sites II and I respectively) have been detected in green plant and algal chloroplasts. Since each site exhibits an $\rightarrow H^+/e^-$ quotient of 1, and the $\rightarrow H^+/ATP$ quotient in chloroplasts is probably 3, the maximum ATP/e^- quotients during cyclic and non-cyclic phosphorylation can be predicted to be approximately 0.3 and 0.7 respectively.

Bacteriorhodopsin-dependent photosynthesis

The halobacteria (e.g. *Halobacterium halobium*) are obligately aerobic, extreme halophiles whose natural habitats are salt-flats or stagnant salt lakes. When cultured under conditions of excess oxygen, these organisms contain a membrane-bound respiratory chain which ejects H^+ and is sensitive to cyanide. In contrast, their growth under the more physiological conditions of low oxygen tension and strong illumination is characterized by the appearance of purple patches (0.5 to 1.0 μm in diameter) which cover up to half of the surface area of the plasma membrane; such cells exhibit light-dependent H^+ ejection which is insensitive to cyanide. The *Halobacteriaceae* thus resemble the *Rhodospirillaceae*, some

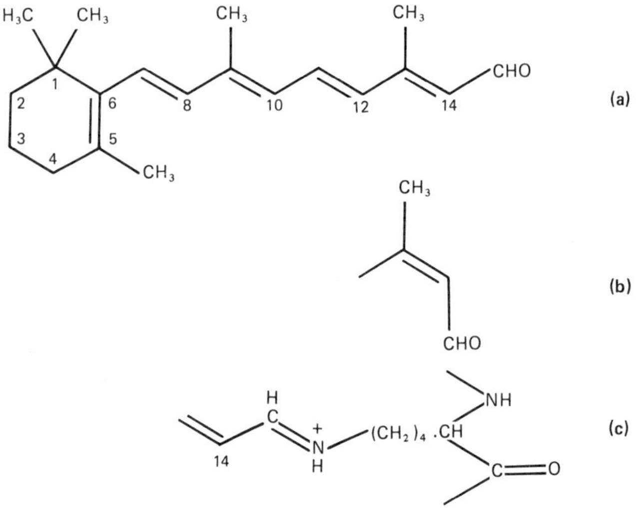

Bacterial Respiration and Photosynthesis

Fig. 4.9 Structural aspects of bacteriorhodopsin. (a) all-*trans* retinal, (b) 13-*cis* retinal, and (c) the Schiff base linkage between retinal and a lysyl residue in bacterio-opsin.

Cyanobacteriaceae, and possibly the *Chloroflexaceae*, in using respiration and photosynthetic electron transfer as alternative methods of conserving energy. In contrast to these latter three families, however, their photosystem contains no bacteriochlorophyll or conventional redox carriers. Carotenoids are present, albeit unusual C_{50} *bacterioruberins*, but these probably serve to protect the cells from the potentially deleterious effects of excess radiation rather than to act as accessory light-harvesting pigments.

Light is absorbed by the purple membrane which contains a pigmented protein, bacteriorhodopsin (bR). The latter consists of retinal (vitamin A aldehyde; Fig. 4.9) covalently bound by a Schiff base linkage to a lysyl residue in the apoprotein bacterio-opsin (MW 25 500); it thus has a remarkable structural resemblance to the pigment rhodopsin, or visual purple, that functions in vertebrate vision. In the dark the retinal moiety exists as a mixture of the 13-*cis* and all-*trans* isomers, forming a dark-adapted purple complex with bacterio-opsin, and exhibiting an absorption maximum at 560 nm (bR_{560}). The results of investigations using low temperature and rapid-flash spectroscopy indicate that low intensity light causes bR_{560} to undergo a 13-*cis* → all-*trans* isomerization to form a light-adapted purple complex (bR_{570}) which is subsequently converted, via a high light intensity photoreaction and a series of dark reactions, into a bleached complex (bR_{412}); the latter is finally reconverted to bR_{570} via further dark reactions. The photo-reaction is very much faster than any of the subsequent dark reactions ($t_{\frac{1}{2}} \leq 10$ ps compared with 1 μs to 5 ms). Since the Schiff base is protonated in bR_{570} ($-\overset{H}{C}=\overset{+}{\underset{H}{N}}-$) but not in bM_{412} ($-\overset{H}{C}=N-$), the cyclical interconversion of these two complexes is accompanied by the sequential release and uptake of a proton. There is some evidence that the bleached complex contains 13-*cis* retinal, thus suggesting that an

Photosynthesis

all-*trans* → 13-*cis* → all-*trans* isomerization sequence may occur during the reaction cycle, but the detailed chemistry of the cycle intermediates is unknown. Physical studies have indicated that each molecule of bacteriorhodopsin contains 7 α-helical regions that are oriented perpendicular to the plane of the membrane such that they zig-zag across almost its entire width. Furthermore, there is evidence that the bacteriorhodopsin encompasses a hydrophilic, proton-translocating channel that spans the membrane and which has a central hydrophobic core that contains the retinal chromophore. It is envisaged that illumination causes a conformational change in the protonated bacteriorhodopsin such that the proton is released into the outer channel and is thence ferried to the periplasm via a series of low pK groups (e.g. carboxyl). The bleached, deprotonated chromophore subsequently receives a proton from the cytoplasm, via a series of high pK groups (e.g. lysyl) that comprise the inner channel, and the bacteriorhodopsin returns to its original conformation (Fig. 4.10). Alternatively it is possible that these acidic/basic groups are replaced by a transmembrane 'ice' channel that allows proton movement to occur via H$^+$ tunnelling along its hydrogen-bonded lattice structure.

Essentially homogeneous populations of right-side-out membrane vesicles (resealed cell envelopes) have recently been prepared from whole cells of

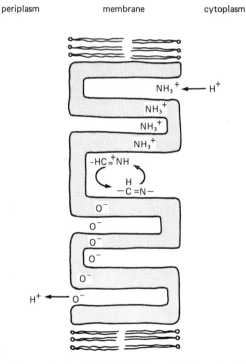

Fig. 4.10 A possible mechanism for the light-dependent translocation of protons via bacteriorhodopsin (after Kozlov & Schulachev, 1977). -NH$_3$, -amino groups of lysyl residues; -O, carboxylate groups of acidic amino acids, and phosphate or sulphate groups of phospholipids and sulpholipids.

H. halobium. In addition, the ease with which fragments of purple membrane can be purified (by differential and density gradient centrifugation following the ready lysis of whole cells at low salt concentrations, e.g. 0.15 M NaCl) has considerably facilitated the reconstitution of artificial membrane vesicles from phospholipids and purple membrane (BR-proteoliposomes). Like whole cells, both types of vesicle exhibit light-dependent H^+ translocation (outwards in right-side-out vesicles, and in either direction in bR-proteoliposomes depending on the conditions used for their reconstitution). The effects of various ionophores on these light-induced H^+ movements confirm that, as in other bacteria, both ΔpH and $\Delta \psi$ contribute to Δp; under static head conditions the latter is in the range 172 to 230 mV. These values are compatible with the major functions of the Δp in *H. halobium* which are (i) to drive ATP synthesis and motility, (ii) to maintain the necessary salt balances demanded by the halophilic nature of the organism (1.3 M Na^+ and 3 M K^+ inside compared with approximately 4 M Na^+ and 0.03 M K^+ outside), and (iii) to power the uptake, either directly or indirectly, of the amino acids which constitute the preferred carbon substrates for growth. Illumination of whole cells of *H. halobium* causes inhibition of respiration, presumably as a result of back-pressure from the light-induced Δp because no inhibition is observed in the presence of uncoupling agents.

There is increasing evidence that the H^+-pump is accompanied by a light-dependent Na^+ pump that is located in the residual red membrane rather than in the purple patches. This pump, which is probably the property of a retinal-based chromoprotein similar to bacteriorhodopsin, generates an electrochemical potential difference of Na^+ ($\Delta \bar{\mu}_{Na^+}$; analogous to $\Delta \bar{\mu}_{H^+}$ or Δp) that can be used to drive solute transport.

The quantum requirement for the ejection of one proton by *H. halobium* is approximately 2 compared with 0.5 in the purple bacteria. Thus, since 570 nm light has a higher energy content than light at 870 nm, the overall efficiency of energy conservation in *H. halobium* is less than one quarter of that in the *Rhodospirillaceae*. This low efficiency is not helped by the apparent inability of the organism to invaginate its plasma membrane, which means that it exhibits a relatively small area of light-absorbing pigment per cell compared with other photosynthetic bacteria. However, these deficiencies are partly offset by the savings in biosynthetic resources which accrue from having a light-dependent H^+ (and Na^+) translocation system which is composed of only one or two proteins, rather than the multiprotein complexes that characterize other phototrophs, and which occupies a convenient niche in the visible spectrum (envisage an absorption band at 560 to 570 nm in Fig. 4.2).

Summary

Photosynthetic energy conservation in bacteria may be classified into three types according to whether the major light-harvesting pigment is bacteriochlorophyll, chlorophyll or bacteriorhodopsin.

During bacteriochlorophyll- and chlorophyll-dependent photosynthesis, which are found respectively in purple or green bacteria and in blue-green bacteria, long wavelength light energy is absorbed by the antennae photopigments and is hence transferred to the reaction centre complex, where it causes a charge separation that produces a low redox potential reductant (Q^- Fe or reduced Fe-S protein) and a high redox potential oxidant (ferricytochrome c or oxidized plastocyanin); electron transfer from the former to the latter is effected via the quinone-cytochrome system and is accompanied by electrogenic proton translocation. This cyclic electron transfer is accompanied by non-cyclic electron transfer from a high redox potential exogenous reductant to $NAD(P)^+$. The reduction of NAD^+ by succinate in some purple bacteria is driven by the proton gradient generated by cyclic electron transfer, whereas the reduction of NAD^+ by reduced sulphur compounds in green bacteria is effected directly via the photosystem and light. Non-cyclic electron transfer from water to $NADP^+$ in blue-green bacteria involves an additional shorter wavelength photosystem (PSII), which contains novel phycobiliproteins as its major light-harvesting pigments, and which oxidizes water concomitant with the release of oxygen and the reduction of plastoquinone; subsequent electron transfer to $NADP^+$ occurs via the cytochrome chain and the longer wavelength photosystem (PSI).

Bacteriorhodopsin-dependent photosynthesis is restricted to the purple membrane of halobacteria, and is completely independent of redox carriers. Energy conservation is effected solely by bacteriorhodopsin, which contains retinal (vitamin A aldehyde) covalently bound to bacterio-opsin via a Schiff base linkage that can exist in a protonated (bR_{570}) or deprotonated (bR_{412}) state. The cyclical, light-dependent interconversion of these two states is accompanied by the assymetric release and uptake of protons across the coupling membrane.

Whole cells, chromatophores and right-side-out vesicles of photosynthetic bacteria are able to convert light energy into a bulk-phase Δp of the required magnitude and direction to drive ATP synthesis and other energy-dependent membrane reactions *in vitro*.

References

CLAYTON, R. K. and SISTROM, W. R. (1978). *The Photosynthetic Bacteria* Plenum press, New York and London.

DREWS, G. (1978). Structure and development of the membrane system of photosynthetic bacteria. *Current Topics in Bioenergetics* 8: 161–207.

DUTTON. L. P. and PRINCE, R. C. (1978). Energy conversion processes in bacterial photosynthesis. In: *The Bacteria*. Vol 6. pp. 523–84. Edited by L. N. Ornston and J. R. Sokatch. Academic Press, New York and London.

EISENBACH, M. and CAPLAN, S. R. (1979). The light-driven proton pump of *Halobacterium halobium*; mechanism and function. *Current Topics in Membranes and Transport* 12: 166–240.

JONES, O. T. G. (1977). Electron transport and ATP synthesis in the photosynthetic bacteria. In: *Mictobial Energetics* pp. 151–83. Edited by B. A. Haddock and W. A. Hamilton. Society for General Microbiology Symposium 27. Cambridge University Press, Cambridge.

KE, B. (1978). The primary electron acceptors in green plant photosystem 1 and photosynthetic bacteria. *Current Topics in Bioenergetics* 7:

MARRS, B. L. (1978). Mutations and genetic manipulations as probes of bacterial photosynthesis. *Current Topics in Bioenergetics* 8: 261–94.

SAUNDERS, V. A. (1978). Genetics of *Rhodospirillaceae*. *Microbiological Reviews* 42: 357–84.

5 Energy transduction

The previous three chapters have described the various types of redox systems and the bacteriorhodopsin-linked H^+ pump which different species of bacteria use to conserve chemical or electromagnetic energy in the form of a trans- or intramembrane proton gradient. Although the latter can be used to drive a variety of physiologically important membrane functions (including certain types of solute transport, reversed electron transfer and motility), its major use is to power the thermodynamically unfavourable synthesis of ATP from ADP and inorganic phosphate (and, in some species of phototrophic bacteria, the synthesis of inorganic pyrophosphate from orthophosphate). In contrast, many fermentative anaerobes with no electron transfer or H^+ pumping capabilities hydrolyse ATP previously synthesized via substrate-level phosphorylation to drive membrane energization. The transduction of energy from a membrane-associated proton/charge gradient (e.g. Δp) into the free energy of hydrolysis of ATP within the cytoplasm (the phosphorylation potential or phosphate potential, ΔGp) or vice-versa, is catalysed by the membrane-bound ATP phosphohydrolase.

The ATP phosphohydrolase ($BF_0 \cdot BF_1$)

ATP phosphohydrolases have recently been investigated from a variety of bacteria which utilize different methods of energy conservation, most prominently *Clostridium pasteurianum* and *S. faecalis* (substrate-level phosphorylation), *E. coli*, *S. typhimurium*, *M. lysodeikticus*, *M. phlei* and the thermophile PS3 (oxidative phosphorylation), and *R. rubrum* and *Chromatium vinosum* (photophosphorylation). As yet, no extensive studies have been reported using groups of bacteria like the halophiles, acidophiles or alkaliphiles which have well-defined energetic problems imposed by their growth environments.

They all consist of two oligomeric proteins (BF_0 and BF_1), and thus resemble the ATP phosphohydrolases of mitochondria ($F_0 \cdot F_1$) and chloroplasts ($CF_0 \cdot CF_1$). BF_1 is an assymetrically-located hydrophilic protein assembly which is responsible for the dual catalytic functions of the overall complex; it is easily detached from the coupling membrane, but in its soluble form catalyzes only ATP hydrolysis. The latter is generally inhibited by a variety of classical ATPase inhibitors including 4-chloro-7-nitrobenzofurazan (Nbf-Cl), aurovertin, azide and quercetin; efrapeptin inhibits only BF_1 from photosynthetic bacteria and *C. pasteurianum*. In contrast, BF_0 is a hydrophobic lipoprotein complex which forms an intrinsic part of the coupling membrane and facilitates the passage of H^+ to and from the active site of $BF_1 \cdot BF_0$ is invariably sensitive to N, N'-dicyclohexyl-carbodiimide (DCCD), and BF_0 from *C. pasteurianum* and various photosynthetic bacteria are inhibited by butyricin and oligomycin respectively.

The energy-transducing ATP phosphohydrolases of bacteria thus grossly resemble other membrane-bound ion pumps (e.g. the H^+, Ca^{2+} and Na^+/K^+

ATPases of higher organisms) in that they are comprised of a gated pump (BF_1) and a channel filter (BF_0) which together are responsible for the catalysis, regulation and selectivity of H^+ translocation.

The isolation, subunit composition and function of BF_1 BF_1 is readily released from bacterial coupling membranes either by washing the membranes several times in Mg^{2+}-free low ionic-strength buffers (e.g. 2mM Tris-HCl), by briefly sonicating them in the presence of a suitable metal chelating agent (e.g. 1 mM EDTA), or by exposing them to selected organic solvents (e.g. *n*-butanol), detergents (e.g. Triton X-100, sodium dodecyl sulphate) or chaotropic agents (e.g. guanidine-HCl, urea). The solublized enzyme is then purified to homogeneity using standard procedures.

Purified BF_1 is a relatively large protein (MW 350 000 to 410 000) which is similar in size to F_1 and CF_1. Many BF_1 preparations contain tightly-bound inorganic phosphate and adenine nucleotides; distinct binding sites for ADP and ATP appear to be present, and enzymic activity requires Mg^{2+} or Ca^{2+}. All of the purified BF_1 preparations examined so far exhibit certain properties which are quite different to those shown by their membrane-bound $BF_0.BF_1$ counterparts. Thus, in addition to their insensitivities to certain $BF_0.BF_1$ inhibitors (e.g. DCCD and, where appropriate, oligomycin and butyricin), most BF_1 preparations exhibit cold-lability, a property which is usually associated with the dissociation of a large multimeric protein into its constituent subunits. This phenomenon, in which a solubilized enzyme exhibits different properties compared with its membrane-bound form, is called allotopy and is apparently common to all energy-transducing ATP phosphohydrolases.

Analysis of dissociated BF_1 using polyacrylamide gel electrophoresis in the presence of sodium dodecyl sulphate (SDS-PAGE) generally indicates the presence of five non-identical subunits which are termed α, β, γ, δ, and ε (MW approximately 56 000, 52 000, 32 000, 13 to 21 000 and 7 to 16 000 respectively). BF_1 preparations from some Gram positive aerobes appear to lack either the δ or ε subunit, but since it is known that the δ subunit content of *E. coli* BF_1 varies with the precise nature of the isolation-purification procedure, it is possible that these small peptides are present *in vivo*. BF_1 from vegetative cells of the obligate anaerobe *C. pasteurianum* appears to contain only three types of subunit (MW 65 000. 57 000 and 43 000); BF_1 from sporulating mother cells of the same organism contains an additional subunit (MW 10 000; probably ε). Analysis of relative subunit concentrations indicates that the five subunit BF_1 probably has the composition $\alpha_3\beta_3\gamma\delta\varepsilon$, but since the δ and ε peptides comprise only approximately 10% by weight of the total enzyme this stoichiometry should be treated with some caution.

The extremely difficult problem of determining the functions of the individual subunits of BF_1 has been facilitated by the recent advent of reliable methods for physico-chemically manipulating purified and membrane-bound BF_1, and for assaying the energy transduction properties of the enzyme in its natural and altered states. Physico-chemical manipulations include (i) the selective removal and large scale purification of the individual subunits of BF_1, (ii) the partial or complete reconstitution of BF_1 from these subunits, (iii) the preparation of inside-out membrane vesicles which either contain BF_1 or are stripped of BF_1, and (iv) the artificial reconstitution of $bR.BF_0.BF_1$ or $BF_0.BF_1$ proteoliposomes from phospholipids, bacteriorhodopsin, BF_1 (intact or partially reconstituted) and BF_0. Energy transduction properties are generally investigated, in the presence or absence of putative inhibitory agents, by measuring (i) ATP hydrolysis, (ii) ATP

Bacterial Respiration and Photosynthesis

synthesis at the expense of Δp generated by respiration, light or artificial means, (iii) ATP- or respiration-dependent membrane energization (e.g. reversed electron transfer or the uptake of selected solutes) and (iv) ATP-dependent proton translocation.

The results of there investigations, principally using BF_1 from *E. coli* and the thermophile PS3, generally indicate that the α and β subunits form the active site of BF_1, possibly in association with the γ subunit. This conclusion is based on the observations that (i) antibodies against $\alpha\beta$ subunits inhibit ATP hydrolysis and ATP-dependent membrane energization, (ii) the proteolytic enzyme trypsin stimulates the ATPase activity of intact BF_1 (from *E. coli*, but not PS3) whilst at the same time digesting significant amounts of the smaller subunits but leaving the α and β subunits largely unimpaired, (iii) mixtures of $\beta\gamma$, $\alpha\beta\gamma$ and $\alpha\beta\delta$ subunits exhibit substantial ATPase activity (as also do mixtures of $\alpha\beta$ subunits from *E. coli* but not PS3), but combinations of subunits which do not include the β subunit are inactive. The isolated β subunit has no ATPase activity *per se*; activity is only attained in association with α and/or γ subunits which presumably endow the complex with the necessary conformational properties. The β and α subunits probably contain the catalytic site and an allosteric site respectively. Studies with [^{3}H]-Nbf-Cl indicate that it binds to the β subunit, only one mole Nbf-Cl per mole of enzyme being required for complete inhibition of ATPase activity.

The relatively basic γ subunit appears to promote the successful assembly of the α and β subunits into a functional catalytic unit, probably through ionic bonding with the relatively acidic β subunits. Its major function, however, may be to control the passage of H^+ to and from the active site of BF_1, since the absence of the γ subunit from partially reconstituted $BF_0 \cdot BF_1$ proteoliposomes of the thermophile PS3 and from selected ATP phosphohydrolase mutants eliminates ATP synthesis and ATP-dependent membrane energization, but not ATP hydrolysis. The γ subunit probably contains the azide-binding site.

There is now reasonably good evidence that the δ and ε subunits facilitate the binding of BF_1 to BF_0. This conclusion is based on the observation that only BF_1 preparations which contain both the δ and ε subunits can bind to BF_1-depleted membranes or $BF_0 \cdot$ proteoliposomes and thus reconstitute ATP-dependent membrane energization. It is likely therefore that these two subunits comprise the proton-translocating stalk which connects BF_1 to BF_0; together with the γ subunit they probably form the gate which regulates H^+ movement between the pump ($\alpha_3\beta_3$ or $\alpha_3\beta_3\gamma$) and the membrane channel (BF_0). The involvement of the ε subunit in the H^+ gate is supported by the observations that the ATPase activity of BF_1 from many organisms is stimulated by trypsin (which readily destroys the ε subunit), and that the purified ε subunit potently inhibits the ATPase activity of complete and ε-deficient BF_1, but not of the membrane-bound enzyme. However, since BF_1 from the thermophile PS3 is inhibited by trypsin and unaffected by supplements of the ε subunit, the exact role of the latter appears to be species-dependent.

The isolation, subunit composition and function of BF_0

BF_0 has no readily assayable enzyme activity. It is preferable therefore to isolate $BF_0 \cdot BF_1$ rather than BF_0 alone, since the former can be assayed via the sensitivity of its ATPase activity to DCCD or other inhibitors of BF_0. Active $BF_0 \cdot BF_1$ complexes have recently been isolated from *E. coli*, thermophile PS3, *R. rubrum* and *C. pasteurianum* following exposure of coupling membranes to selected detergents such as cholate, deoxycholate or Triton X-100.

Analysis of these $BF_0.BF_1$ complexes using SDS-PAGE indicates that the peptide components of BF_0 are strongly hydrophobic and vary widely in number according to origin, although much of this variation may reflect incomplete purification of the $BF_0.BF_1$ complex. All of the BF_0 complexes so far examined contain a DCCD-binding proteolipid (η, MW 6 500 to 15 000); indeed the BF_0 of *C. pasteurianum* seems to be composed solely of η subunits (possibly η_8) which are therefore responsible for binding DCCD, butyricin and BF_1. In contrast, a separate BF_1-binding subunit (ζ, MW 13 500) has been isolated from the thermophile PS3, and these two subunits appear to comprise the BF_0 of this organism, probably as $\zeta_3\eta_6$. Although at least one additional peptide (MW 24 000 to 28 000) is present in BF_0 from *E. coli* and *R. rubrum*, it would appear that the simpler function of BF_0 compared with BF_1 (i.e. that of an H^+ channel instead of an H^+ gate and pump) is reflected in its smaller size and lesser complexity.

The DCCD-binding subunit of *E. coli* has also been purified without the intermediate isolation of the $BF_0.BF_1$ complex following extraction of [^{14}C]DCCD-labelled membranes with chloroform-methanol. The [^{14}C]-labelled component is a proteolipid (MW 8 000) which is identical with the η subunit, the role of which is probably to form a transmembrane channel that normally allows H^+ and water to pass to and from BF_1. This conclusion stems mainly from the observations that (i) ATP synthesis at the expense of light, respiration or an artificially-imposed Δp is inhibited by DCCD, as also is the generation of Δp at the expense of ATP hydrolysis, (ii) incorporation of either native BF_0 or reconstituted BF_0 into liposomes in the absence of DCCD usually renders them specifically permeable to H^+, i.e. passive H^+ translocation occurs in the presence of a suitable Δp, (iii) the H^+ permeability of such proteoliposomes is abolished by the addition of BF_1 or antibodies to BF_0, but not by the addition of δ and/or ε subunits unless supplemented by the γ subunit (thus confirming the gating action of the $\gamma\delta\varepsilon$ complex), and (iv) DCCD repairs the loss of respiration-linked energy transduction, caused by increased H^+ permeability, which characterizes membrane vesicles prepared from appropriate Unc^- mutants of *E. coli* (see below) or which results from the exposure to chaotropic agents of membrane vesicles from wild type cells.

In view of the protonophoric properties of the η subunit, it is interesting to note that in its hexameric or octameric state it is considerably larger than channel-forming ionophores such as gramicidin, thus making a transmembrane orientation perfectly feasible. The pH profile for proton translocation through BF_0-proteoliposomes confirms that H^+ rather than OH^- is the translocated species, and indicates the presence of monoprotic binding sites in BF_0 which are compatible with H^+ transfer via a chain of adjacent histidyl residues.

Subunit organization in $BF_0.BF_1$ The elegant reconstitution experiments described above throw considerable light not only on the functions of the individual subunits of the ATP phosphohydrolase, but also on their location and proximity within the complex. Further information about the subunit organization of BF_1 has been obtained by computer filtering of optical diffraction patterns from electron micrographs, and by using various reagents which cross-link adjacent polypeptides through their functional groups. The results suggest that the α and β subunits are arranged alternately to form a hollow planar hexagon, the central hole of which contains the γ subunit. The observation that some BF_1 preparations which are deficient in the δ and/or ε subunits are nevertheless catalytically active, confirms

Bacterial Respiration and Photosynthesis

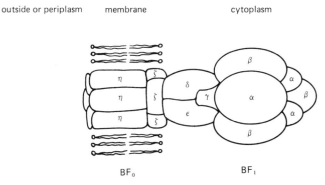

Fig. 5.1 A diagrammatic representation of the subunit organization in $BF_0 \cdot BF_1$. The entire complex is composed of an H$^+$ pump ($\alpha_3\beta_3$ or $\alpha_3\beta_3\gamma$), an H$^+$ gate ($\gamma\delta\varepsilon$), and an H$^+$ channel through the coupling membrane ($\zeta_3\eta_6$). It should be noted that this model does not accommodate the claim that the three α subunits in $BF_0 \cdot BF_1$ from *S. faecalis* may contain protein 'tails' which assist the stalk subunits in anchoring BF_1 to the membrane. Furthermore, $BF_0 \cdot BF_1$ from *C. pasteurianum* lacks the δ and ζ subunits, and BF_0 from *E. coli* and *R. rubrum* contains at least one additional subunit.

that the δ and ε subunits are not involved in binding the γ subunit to the $\alpha_3\beta_3$ hexagon. This conclusion is supported by convincing evidence that the principal functions of the δ and ε subunits, in association with the γ subunit, are to bind the H$^+$ pump to the ζ subunits of BF_0, and to facilitate and regulate the reversible transfer of protons between the DCCD-binding channel in BF_0 and the active site of BF_1 (Fig. 5.1).

$BF_0 \cdot BF_1$: ATPase or ATP synthetase?

There is now considerable evidence that the function of the ATP phosphohydrolase varies with the type of energy conservation imposed by the physiology of the organism and/or by the growth environment. During strictly fermentative growth under anaerobic conditions, ATP is synthesized directly by the various macroscopic and scalar chemical reactions which are characteristic of substrate level phosphorylation. Although the majority of the ATP so formed is subsequently used as an immediate source of energy for biosynthetic purposes, a significant proportion undergoes controlled hydrolysis leading to the generation of the membrane-associated proton/charge gradient which is responsible for driving reversed electron transfer, motility and certain forms of solute transport (and which can probably be enhanced by proton ejection in symport with anionic fermentation products such as lactate or acetate). In contrast, during growth in the presence of light and/or an appropriate oxidant the energy released by the vectorial reactions of bacteriorhodopsin, photosynthetic electron transfer or respiration is conserved initially as the proton/charge gradient, and only subsequently is the latter used to drive ATP synthesis. Thus, in the former situation $BF_0 \cdot BF_1$ acts as a proton-ejecting ATPase, whereas in the latter case it functions as a proton-injecting ATP synthetase.

Energy transduction

The physiological role of the ATP phosphohydrolase complex *in vivo* is reflected in its properties *in vitro*. Thus, the enzyme from vegetative cells of the fermentative anaerobe *C. pasteurianum* has a high ATPase activity, the Km (ATP) for which is lowered by intermediates or products of fermentation; ATP synthetase activity is low and susceptible to product inhibition. In contrast, the ATP phosphohydrolases of bacteria grown on non-fermentable substrates exhibit high ATP synthetase activities which are not readily inhibited by ATP, and their ATPase activities are often low until activated by controlled proteolysis of the ε subunit. It is tempting to correlate the absence of the ε subunit with high ATPase activities, and to speculate that in many non-fermentative organisms the role of this subunit (possibly in association with other subunits) is normally to direct the ATP phosphohydrolase in favour of ATP synthesis, via an as yet undetermined mechanism. Indeed, it is possible that the four-subunit ATP phosphohydrolase of the modern *C. pasteurianum* is a closely-related descendant of a primitive hydrolytic form of the enzyme which, during the evolution of bacteria capable of photosynthesis or respiration, acquired a further two or more subunits ($\zeta\delta$) in order to facilitate the rapid and efficient synthesis of ATP. The fact that the fermentative anaerobe *S. faecalis* has a complete five-subunit BF_1 is by no means incompatible with this hypothesis since this organism may be regarded as a degenerate aerobe, i.e. it will grow aerobically in the presence of haematin, under which conditions it conserves energy by oxidative phosphorylation.

Energy transduction mutants

Investigations into the functions of $BF_0.BF_1$ have recently received considerable impetus by the isolation of mutant strains of *E. coli* which are defective in energy transduction (Unc$^-$). These mutants (i) grow aerobically on glucose but not on non-fermentable carbon sources such as succinate or malate, (ii) exhibit aerobic molar growth yields on glucose which are similar to those of the wild type strains growing anaerobically, and (iii) grow anaerobically on glucose only under conditions where anaerobic respiration is possible. These growth patterns indicate defects in the ATP phosphohydrolase which cause the mutants to be energetically uncoupled (i.e. they are incapable of oxidative phosphorylation and ATP-dependent membrane energization, but their capacity for respiration is unimpaired and they have some ability to utilize the energized protons generated by electron transfer). These properties thus explain why both a fermentable carbon source and a terminal electron acceptor are obligatory for growth, since the former provides, via substrate level phosphorylation, the ATP which is necessary for the biosynthesis of cell components, whilst the latter furnishes, via respiration, the proton/charge gradient which effects membrane energization and hence is responsible for driving reversed electron transfer, motility and some types of solute transport (Fig. 5.2). Since some of these mutants exhibit ATPase activity, whilst others do not, this criterion can be used to divide them into two distinct classes: Unc$^-$ ATPase$^-$ and Unc$^-$ ATPase$^+$.

A large number of mutants of both classes have now been isolated, but many of them are as yet incompletely characterized. This discussion is therefore restricted to several which have been studied extensively and which exhibit significantly different lesions.

Bacterial Respiration and Photosynthesis

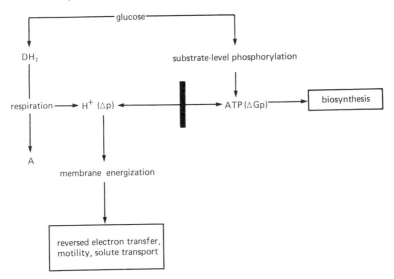

Fig. 5.2 A schematic view of cellular energy transduction showing the requirement for both respiration and substrate-level phosphorylation during the growth of Unc⁻ mutants of *E. coli*. Respiration liberates the protons which are responsible for membrane energization, whilst substrate level phosphorylation provides the ATP which is required for biosynthesis. The heavy line indicates genetic lesions in BF_0, BF_1 which prevent Unc⁻ mutants from interconverting Δp and ΔGp.

Unc⁻ ATPase⁻ (*unc A*, *unc D*, DL 54) The ATPase⁻ phenotype can obviously reflect various defects, most of which are associated with BF_1, e.g. (i) failure to synthesize BF_1, (ii) synthesis of BF_1 with a defective catalytic ability, or (iii) synthesis of BF_1 with an impaired ability to bind to the coupling membrane. The *unc A* mutant appears to fall into the second of these categories, since its BF_1 is membrane-bound but exhibits no ATPase activity. Furthermore, the unique ability of a hybrid ATPase complex, constructed from wild type BF_1 plus BF_1-depleted membranes from the *unc A* mutant, to exhibit oxidative phosphorylation and various ATP-dependent membrane reactions confirms that the *unc A* lesion is located in BF_1 rather than BF_0; the normal proton conductance and respiratory chain energization properties of the mutant support this conclusion. Complementation experiments using purified subunits from *unc A* and wild-type BF_1 indicate that the lesion leads to an abnormal α subunit. The *unc D* mutant has similar properties to *unc A* but is characterized by an abnormal β subunit.

In contrast, BF_1 in mutant DL 54 is only poorly bound to the coupling membrane. The latter contains many unfilled binding sites, has relatively poor respiration-dependent energization properties and exhibits a high permeability to protons; these last two defects can be repaired by the addition of DCCD or wild type BF_1. It is likely, therefore, that BF_1 in this mutant fails to plug completely the proton translocating channels through BF_0 and thus impairs the ability of the coupling membrane to maintain a Δp; the lesion is probably in the δ or ε subunit.

Energy transduction

Unc⁻ATPase⁺ (*unc B, unc C*) Apart from their ability to catalyze ATP hydrolysis, and the tendency of the latter in some mutants to be resistant to inhibition by DCCD, the energy transduction properties of membrane vesicles prepared from Unc⁻ATPase⁺ mutants are not grossly dissimilar to those prepared from Unc⁻ATPase⁻ mutants. These properties indicate defects in those components of the ATP phosphohydrolase complex which are responsible either for binding BF_1 to BF_0 or for regulating the movement of protons through the coupling membrane. Analyses of hybrid ATPase complexes reconstituted from BF_1 and BF_1-depleted membranes from mutant and wild type strains indicate that the Unc⁻ATPase⁺ phenotype can reflect alterations in either BF_0 or BF_1. Thus *unc B* mutants are resistant to DCCD and contain a defect in a membrane component which is associated with the H^+-translocating channel, whereas *unc C* mutants exhibit DCCD sensitivity and are defective at the $BF_0.BF_1$ interface. Since these lesions lead to a complete loss of energy transduction at the expense of ATP hydrolysis, but not at the expense of respiration, it would appear that only the respiratory chain (and not the $BF_0.BF_1$ complex) of such mutants can translocate protons fast enough to generate a significant Δp; even then the latter is apparently sufficient only to drive active transport or transhydrogenation, but not the net synthesis of ATP. The failure of the ATP phosphohydrolase complex to generate or utilize Δp reflects quite different membrane properties in these two mutants, i.e. an increased permeability to H^+ in *unc C* (defective δ, ε or ζ subunit.) and an inability to form an effective H^+-translocating channel in *unc B* (defective η subunit).

Inhibitor-resistant mutants Such mutants are selected by their ability to grow on solid medium containing a non-fermentable carbon source such as succinate, malate or acetate supplemented with normally lethal concentrations of an ATP phosphohydrolase inhibitor. In this way mutants have been isolated which are resistant to DCCD (*E. coli* RF-7 and DC1, *S. faecalis* dcc-8) and aurovertin (*E. coli* MA1, MA2). Phenotypically, such mutants exhibit energy transduction properties comparable to those of the wild type organisms, but their ATPase and ATP-dependent membrane energization reactions are very much less sensitive to the appropriate inhibitor.

Hybrid ATPase complexes reconstituted from mixtures of BF_1 and BF_1-depleted membranes prepared from wild type and DCCD-resistant mutants exhibit properties which show unambiguously that the locus of DCCD-inhibition is BF_0 rather than BF_1, thus supporting its assignation to the η subunit. SDS-PAGE analysis of membranes prepared from these mutants generally indicates the continued presence of this subunit, although the latter is presumably modified in such a way that many of its potentially reactive functional groups (in particular its several carboxyl groups) are no longer accessible to DCCD. In contrast, aurovertin resistance is characteristic of BF_1 but the subunit responsible has not yet been unambiguously identified.

Uncoupler-resistant mutants have also been isolated (*B. megaterium* C8 and C12). These mutants exhibit similar energy transduction properties to those of wild type cells except that oxidative phosphorylation, but not amino acid transport, is insensitive to uncoupling agents. Following extensive analysis of this phenomenon it has been suggested that uncoupling agents not only render the membrane permeable to H^+ (thus abolishing Δp and inhibiting solute transport), but also bind to a specific site on $BF_0.BF_1$ (hence blocking ATP synthesis); this site is apparently defective in the mutants, which thus appear capable of catalysing Δp-independent

Bacterial Respiration and Photosynthesis

oxidative phosphorylation. It is possible that the latter occurs at the expense of a more localized proton concentration which is not in equilibrium with Δp.

The unc operon All of the Unc$^-$ and inhibitor-resistant mutants of *E.coli* which have so far been examined genetically map at approximately 83 min on the chromosome and are 70% cotransducible with *asn*. The structural genes which code for the polypeptide subunits of BF_0 and BF_1 thus form a relatively tight cluster; indeed, the five *unc* genes which have so far been satisfactorily identified by genetic complementation analysis appear to constitute an operon which is transcribed in the order *unc BEADC* (Fig. 5.3). Since the ATP phosphohydrolase from *E.coli* contains a minimum of eight non-identical polypeptides, it is likely that the *unc* operon contains at least three more as yet incompletely identified structural genes (*unc F*, *unc G* and *unc H*) which are probably located between the *unc A* and *unc D* genes. There is some evidence that the *unc G* gene codes for the γ subunit, but the products of the *unc E*, *unc F* and *unc H* genes are not known. Apart from some evidence that a regulatory gene for this operon may be located at approximately 77 min on the chromosome, virtually nothing is currently known about the mechanisms which regulate the synthesis and organization of the subunits or which control the integration of the completed BF_0 and BF_1 complexes into the coupling membrane.

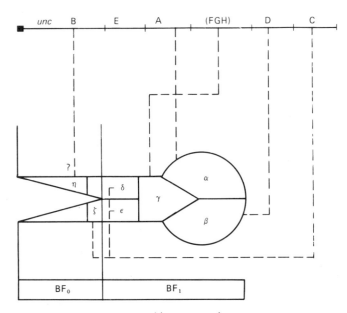

Fig. 5.3 The *Unc* operon and its gene products.

Energy transduction

The energetics of ATP synthesis and hydrolysis

The protonmotive force Considerable experimental attention has recently been focussed on the thermodynamics of ATP synthesis and hydrolysis in bacteria, principally with a view to assessing the competence of chemiosmotic theory to describe energy transduction.

It is now clear that both respiration and photosynthesis can generate a Δp (100 to 420 mV; inside alkaline and electrically negative in whole cells and right-side-out vesicles, but acidic and electrically positive in chromatophores, inside-out vesicles and reconstituted proteoliposomes) to which ΔpH and $\Delta \psi$ contribute variably according to the assay conditions. It is therefore of fundamental importance to ascertain whether these values of Δp are sufficient to drive net phosphorylation, and whether the latter can also be effected by ΔpH or $\Delta \psi$ alone. In order to do this, ATP synthesis is monitored as a function of ΔpH and/or $\Delta \psi$, these different driving forces being generated either physiologically (at the expense of light or respiration, in the presence or absence of ionophorous antibiotics or permeant ions to selectively collapse ΔpH or $\Delta \psi$) or artificially (through rapid acid-base transitions and/or valinomycin-induced K^+ movements to generate ΔpH and $\Delta \psi$ respectively).

The results of such experiments using whole cells, membrane vesicles, chromatophores and $bR.BF_0.BF_1$-or $BF.BF_1$-proteoliposomes indicate that it is generally the magnitude rather than the composition of Δp which appears to be of major importance in driving ATP synthesis. Thus, steady state phosphorylation can be effected at the expense of either $\Delta \psi$, ΔpH or a combination of the two, but net synthesis of ATP occurs only when the total driving force exceeds a threshold value of approximately 150 mV ($\equiv$ a ΔpH alone of ≥ 2.5 pH units). It would appear, therefore, that the Δp values generated physiologically by respiration or photosynthesis are in the main sufficient to drive phosphorylation. As expected, ATP synthesis at the expense of an artificial or bacteriorhodopsin-mediated Δp is abolished by uncoupling agents and inhibitors of the ATP phosphohydrolase, but not by inhibitors of electron transfer; in contrast, ATP synthesis associated with respiration or photosynthetic electron transfer is inhibited by all three types of reagent.

There is some evidence that H^+ movement through the ATP phosphohydrolase is regulated by a pH-dependent and/or voltage-dependent gating mechanism, i.e. conductance is low until Δp exceeds the threshold value for ATP synthesis, at which point conductance increases by several orders of magnitude. This type of regulation, mediated by the $\gamma\delta\varepsilon$ gate complex of BF_1 through an as yet undetermined mechanism, would thus minimize the leakage of H^+ and allow a significant Δp to be generated even at very low rates of H^+ pumping.

Since, at least *in vitro*, ATP synthesis can occur at the expense of a transmembrane Δp, it is of interest to determine the ability of the coupling membrane to generate and store ΔpH and $\Delta \psi$. This property has recently been extensively investigated in chromatophores, where the generation of Δp can be started or stopped instantaneously simply by switching the light on or off; if when the light is switched off the chromatophores are immediately exposed to ADP and inorganic phosphate, the size of the stored Δp is reflected in the extent of ATP synthesis. The results show that following the onset of illumination $\Delta \psi$ and ΔpH are built up at markedly different rates, the former being induced much more rapidly than the latter, as predicted from the known electrical capacitance and buffering properties

Bacterial Respiration and Photosynthesis

of chromatophores. When the light is switched off, the normally low level of post-illumination ATP synthesis is unaffected by the presence of NH_4Cl (which enhances $\Delta\psi$ at the expense of ΔpH), but is dramatically stimulated by SCN^- or by valinomycin + K^+ (which enhance ΔpH at the expense of $\Delta\psi$). It thus appears that $\Delta\psi$ decays much more rapidly than ΔpH, indicating that the chromatophore membrane is highly permeable to selected cations and/or anions but not to protons. The failure of untreated chromatophores to yield high levels of post-illumination ATP presumably reflects the fact that under the conditions employed ΔpH and $\Delta\psi$ contribute approximately equally to Δp, and that the rapid decay of the $\Delta\psi$ leaves a ΔpH which is well below the required threshold value for ATP synthesis.

During fermentative growth the major function of the ATP phosphohydrolase is probably to hydrolyse the ATP produced by substrate-level phosphorylation and thus drive membrane energization. Since the latter can also be effected by respiration or photosynthesis, it follows that ATP hydrolysis must be capable of generating a Δp which is similar in magnitude and direction to that produced via electron transfer or by the action of bacteriorhodopsin. This has recently been confirmed using thermophile $BF_0.BF_1$-proteoliposomes which hydrolyze ATP to produce a $\Delta p \geq 200$ mV. As expected, Δp is effectively abolished by uncoupling agents and ATP phosphohydrolase inhibitors, but not by inhibitors of electron transfer; specific ΔpH- or $\Delta\psi$-collapsing agents do not significantly affect the overall value of Δp, but simply enhance $\Delta\psi$ at the expense of ΔpH, and *vice-versa*.

$\rightarrow H^+/ATP$ **quotients** In order to comply with the ATP/O ($ATP/2e^-$) and $\rightarrow H^+/O(\rightarrow H^+/2e^-)$ quotients which have been determined experimentally in bacteria, a minimum of two protons must be retranslocated through the ATP phosphohydrolase for each molecule of ATP synthesized. The same stoichiometry, but of opposite direction, should therefore characterize ATP hydrolysis.

In theory, $\rightarrow H^+/ATP$ quotients may be determined directly, i.e. kinetically, by measuring either the uptake of H^+ concomitant with ATP synthesis or the ejection of H^+ concomitant with ATP hydrolysis. The first method, which entails measuring changes in endogenous adenine nucleotide concentrations in whole cells, yields $\rightarrow H^+/ATP$ quotients of ≥ 5 g-ion H^+.mole ATP^{-1}; however, these values are probably overestimated due to the uncorrected consumption of ATP by various cellular reactions and to the uncorrected uptake of H^+ via other H^+ transport systems in the cell membrane. The second method, which is based upon measuring the hydrolysis of exogenous ATP, is not feasible with whole cells (since BF_1 is located on the cytoplasmic surface of the coupling membrane and the latter does not contain an adenine nucleotide translocase) and is restricted to inside-out vesicles, chromatophores and $BF_0.BF_1$-proteoliposomes. Furthermore, it is necessary with this method to distinguish between that part of the total pH change which is composed of scalar H^+ production resulting from ionization changes in the reaction components, and that part which reflects vectorial H^+ movement via $BF_0.BF_1$ (Fig. 5.4). The use of this method with inside-out vesicles from *E.coli* indicates a maximum $\rightarrow H^+/ATP$ quotient of approximately 0.6 g-ion H^+.mole ATP^{-1}. However, since this value is not corrected for the inherent leakiness to H^+ of these vesicles, the true $\rightarrow H^+/ATP$ quotient is likely to be considerably higher. Indeed, similar experiments have yielded $\rightarrow H^+/ATP$ quotients of approximately 2 for specially prepared PS3 proteoliposomes which contain only one $BF_0.BF_1$ complex per vesicle (the unusually high phospholipid: protein ratio virtually eliminates H^+ leakage).

Energy transduction

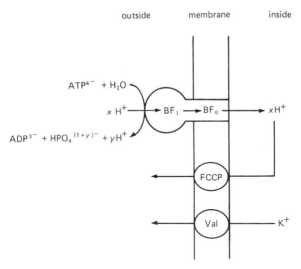

Fig. 5.4 The ATPase reaction in inside-out vesicles, chromatophores and reconstituted proteoliposomes. x is equal to the $\rightarrow H^+/ATP$ quotient and y has a value of up to 0.8/ATP depending on the ambient pH. The rate of ATP hydrolysis is assayed from the continuous acidification of the external medium in the presence of FCCP (to nullify the effect of proton translocation). When FCCP is replaced by valinomycin + K^+, the translocated protons have a greater tendency to remain inside since $\Delta\psi$ is collapsed by the outward movement of K^+; under these conditions the pH changes in the external medium reflect both ionization and translocation. The net rate of H^+ translocation is the rate in the presence of valinomycin + K^+ minus the rate in the presence of FCCP.

$\rightarrow H^+/ATP$ quotients have also been determined indirectly, i.e. thermodynamically, by carrying out parallel assays of Δp and ΔGp, where the latter is defined by the equation:

$$\Delta Gp = -\Delta G^{\theta'} + 2.303\, RT \log \frac{[ATP]}{[ADP][Pi]}$$

in which $\Delta G^{\theta'}$ is the standard free energy change of ATP hydrolysis (kJ.mole^{-1}). The rationale behind this approach is that when the electron transfer or bacteriorhodopsin and ATP phosphohydrolase systems are in equilibrium, Δp is related to ΔGp by the equation:

$$\Delta Gp = \rightarrow H^+/ATP . \Delta p . F$$

When this method is applied to whole cells and inside-out vesicles or chromatophores from selected respiratory and photosynthetic bacteria including *P. denitrificans*, *R. capsulata*, *T. thermophilus* and *M. methylotrophus*, ΔGp is found to vary between 43 and 63 kJ.mole^{-1} ($\Delta Gp/F = 450$ to 650 mV), whereas Δp is usually in the range 150 to 200 mV. The resultant $\rightarrow H^+/ATP$ quotients thus range from approximately 2 to 3 g-ion H^+ .mole ATP^{-1}. However, it should be noted that the Δp values determined by most of these procedures refer to protonic potential differences between the bulk aqueous phases on either side of the coupling membrane. Thus, if the functional proton current occurs not in the bulk phase but in the interphase close to the membrane surface, or even in the membrane itself, it is possible that Δp could be significantly underestimated; indeed, a Δp as high as 420 mV has been obtained with *R. capsulata* chromatophores using the carotenoid band-shift method, which may monitor the surface as well as the bulk-phase $\Delta\psi$. Furthermore, it is unlikely that a completely efficient coupling between H^+ flux and ATP synthesis occurs in isolated vesicles, many of which exhibit significant H^+ leakage.

A kinetic method related to the above, which involves measuring the rate of ATP synthesis as a function of an artificially-imposed bulk phase Δp, yields an $\rightarrow H^+/ATP$ quotient of 1.8 for whole cells of the fermentative anaerobe *Streptococcus lactis*. A similar value has been obtained from short-flash kinetic experiments with *R. sphaeroides* chromatophores which indicate that a single turnover of the reaction centre (i.e. 1 transfer) is sufficient to initiate ATP synthesis. Taken overall, therefore, most data point to an $\rightarrow H^+/ATP$ quotient of approximately 2 g-ion H^+.mole ATP^{-1}. In contrast, $\rightarrow H^+/ATP$ quotients of approximately 3 are now fairly generally accepted for mitochondria and chloroplasts, and such a value would more readily accommodate the high $\rightarrow H^+$/site ratios recently claimed for the respiratory systems of *B. stearothermophilus* and of *P. denitrificans* (detailed analyses of energy transduction in blue-green bacteria should prove interesting in this respect since their photosynthetic properties are similar to those of chloroplasts). It should be noted, however, that the $\rightarrow H^+/ATP$ quotient may not be a universal constant, even within a single organism, and that it may exhibit a range of integral or non-integral values according to the degree of uncoupling, the energy requirements of the cell, and the precise nature of its growth environment.

The mechanism of ATP synthesis

Two important findings have recently thrown some light on the possible enzymic mechanism of ATP synthesis via the ATP phosphohydrolase. These are the observations that only one Nbf-Cl per enzyme is required for full inhibition of activity (even though there is more than one β subunit), and that BF_1 contains at least two relatively weak catalytic binding sites for ATP and ADP (strong binding sites have also been detected and possibly have some sort of regulatory function). From these and other observations it seems increasingly likely that ATP synthesis occurs via an alternating catalytic site mechanism in which there are cooperative interactions between two or more catalytic domains (Fig. 5.5). It is envisaged that membrane energization causes conformational changes in BF_1 that lead to the tighter binding of ADP and phosphate at one site, and to the loosening of ATP binding at another site; ATP is subsequently formed at the first site and the preformed ATP is released from the second site, which now binds ADP and phosphate. This half-cycle is effectively repeated to allow ATP to be released from the first site and to be formed at the second site. It is important to note that energy is not required for the condensation of ADP and phosphate *per se*, but to facilitate the binding of the reactants and the release of the products. However, the way in which the functional proton current is able to induce the required conformational and substrate affinity changes in BF_1 remains to be elucidated.

Energy transduction

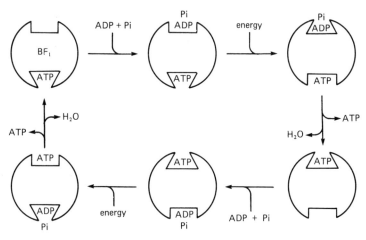

Fig. 5.5 The alternating catalytic site mechanism of ATP synthesis (after Boyer, 1977).

There is increasingly good evidence for short range interactions between various redox systems and the ATP phosphohydrolase which may be indicative of a localized rather than a delocalized proton current, particularly in chromatophores. Thus the steady-state rate of photophosphorylation is by no means a constant function of Δp under all experimental conditions (e.g. in the presence of uncoupling agents, electron transfer inhibitors or different light intensities) and the yield of ATP following a very short flash of light can still be substantial even when the measured Δp is as low as 70 mV. Similarly, detailed analyses of energy transduction in inside-out membrane vesicles from *P. denitrificans* have shown that Δp and ΔGp can vary independently in the presence of certain substrates or permeant ions (whereas the chemiosmotic hypothesis predicts a constant relationship between these two parameters). It seems increasingly likely, therefore, that the bulk-phase transmembrane Δp may imperfectly reflect the actual driving force for ATP synthesis. Furthermore, it is difficult to envisage how alkaliphiles can effect energy transduction via a delocalized proton current, or how many species of bacteria with tightly-packed concentric membrane systems can attain chemiosmotic equilibrium. However, it is known that several organisms arrange parts of their respiratory chains into energy-transducing redox arms (which are compatible with either a localized or a delocalized proton current) and no clear case has yet been reported of bacterial energy conservation in the absence of a defined vesicular structure (i.e. without osmotic properties). It must be concluded, therefore, that neither the chemiosmotic nor the localized proton hypotheses have yet been statisfactorily proved or disproved, although the tide of favour at the moment seems to be flowing perceptibly towards the latter, or at least towards a compromise view in which the proton current is transmembranous but restricted to structured-water or ice channels on the surface of the membrane.

Further aspects of membrane energy transduction

Apart from driving ATP synthesis, the energized coupling membrane powers several other reactions including reversed electron transfer, the synthesis of inorganic pyrophosphate by some members of the *Rhodospirillaceae* family, the transport of certain solutes, and cell motility.

Reversed electron transfer This phenomenon is particularly important to many chemolithotrophic bacteria, which need to reduce NAD^+ at the expense of a higher redox potential reductant such as Fe^{2+} or various nitrogen and sulphur compounds (reversal of sites 2 and/or 1), and to those organisms which reduce $NADP^+$ by an energy-dependent transhydrogenase (reversal of site 0). Net reversal occurs until the Δp generated by forward electron transfer (e.g. at site 3) or ATP hydrolysis is matched by the back pressure of the energy-requiring redox reaction (i.e. $n.\Delta E_h$ divided by the $\rightarrow H^+/2e^-$ quotient of the reaction, where n is the number of electrons transferred and ΔE_h is the difference between the operating redox potentials of the initial and final redox carriers). The utilization of a considerable amount of energy for this process by many chemolithotrophic bacteria is responsible for the particularly low molar growth yields of these organisms.

The synthesis and hydrolysis of inorganic pyrophosphate In addition to their ATP synthetase-ATPase activities, chromatophores from the facultative phototrophs *Rhodospirillum* and *Rhodopseudomonas* catalyse the reversible H^+-translocating synthesis and hydrolysis of inorganic pyrophosphate (PiPi). This reaction:

$$xH^+_{(out)} + PiPi + H_2O \rightleftharpoons 2Pi + xH^+_{(in)} \quad \Delta G^{\theta'} = -21.9 \, kJ.mole^{-1}$$

does not involve the $BF_0.BF_1$ complex since it is not inhibited by DCCD, oligomycin or antibodies to BF_1. The enzyme responsible, pyrophosphate phosphohydrolase, has recently been solubilized and partially purified; it is considerably smaller ($MW \cong 100\,000$) than the ATP phosphohydrolase, is specific for pyrophosphate, and requires both Mg^{2+} and phospholipid for full activity. Pyrophosphate synthesis occurs readily at the expense of light or respiration, particularly in the absence of ADP, and is inhibited by classical electron transfer inhibitors and uncoupling agents. Pyrophosphate hydrolysis generates a Δp (inside acidic and electrically positive) which is sufficient to drive reversed electron transfer and ATP synthesis. Analysis of the pH changes which accompany pyrophosphate hydrolysis indicates an $\rightarrow H^+/PiPi$ quotient of approximately $0.5 \, g$ ion H^+ .mole $PiPi^{-1}$, but this is almost certainly an underestimate.

The ability of these organisms to synthesize and hydrolyse pyrophosphate is probably vestigial of an era, prior to the advent of ATP, when pyrophosphate was an important energy currency molecule. The contemporary role of pyrophosphate is probably to supplement ATP during growth at low light intensities and to act as a precursor of the energy and phosphorus storage polymer, polyphosphate.

Solute transport During the transport of solutes across the cytoplasmic membrane they either undergo chemical modification or remain unaltered. The former (group translocation) is a relatively restricted phenomenon and is best exemplified by the transport of glucose into obligate and facultative anaerobes via the membrane-bound phosphotransferase system, during which process the glucose is phosphorylated to glucose-6-phosphate at the expense of phospho*enol* pyruvate. In

Energy transduction

contrast, unmodified solute transport is extremely common and occurs via three basic mechanisms, i.e. (i) simple diffusion, whereby small uncharged molecules such as hydrogen pass through the membrane without the assistance of carriers; the solute is transported under the driving force of its own concentration gradient, $\Delta\bar{\mu}_S$ ($\equiv Z.\log[S]_{in}/[S]_{out}$; mV), and is not accumulated, (ii) facilitated diffusion, similar to simple diffusion except that specific carriers are involved and the system is saturable by the substrate (e.g. glycerol), and (iii) active transport, whereby charged or uncharged solutes are transported, usually on specific carriers, and are accumulated at the expense of Δp or, less frequently, $\Delta\bar{\mu}_{Na}^{+}$ (e.g. the transport of amino acids, inorganic ions and some sugars, including glucose, in obligate aerobes); in addition, some amino acids are transported at the direct expense of ATP and via the mediation of periplasmic binding proteins.

Extensive investigations of Δp- (or $\Delta\bar{\mu}_{Na}^{+}$)-dependent active transport have shown that the solute carriers are tightly membrane-bound and of varied substrate specificity; they are functionally symmetrical and reversible, catalysing transport in either direction as dictated by the predominant driving force. When Δp initially exceeds $\Delta\bar{\mu}_S$ ($\Delta\psi + \Delta\mu_S$; mV) solute accumulation occurs until a steady-state level is reached; conversely, when $\Delta\bar{\mu}_S$ exceeds Δp solute efflux results. If the latter is accompanied by H^+ and/or charge transfer it can generate a significant ΔpH and/or $\Delta\psi$ (indeed the excretion of lactic acid as a result of homolactic fermentation increases the molar growth yield of *Streptococcus cremoris* by over 30% when the [lactate]$_{out}$ is artificially maintained close to zero such that $\Delta\bar{\mu}_{lactate}$ is high).

The physiological Δp in whole cells (and also in right-side out vesicles; inside alkaline and electrically negative) dictates that the transported species is either neutral or positively charged, i.e. anions and neutral molecules are cotransported with protons (H^+ symport), whereas cations enter either with protons or unaccompanied (cation uniport) and exit in exchange for protons (H^+. cation antiport) (Fig. 5.6). It is clear, therefore, that the active transport of solutes into or out of whole cells occurs at the expense of one or both components of Δp depending on the precise composition of the transported species, e.g. $2H^+$. succinate symport, H^+.Na^+ antiport and $2H^+$.Ca^{2+} antiport (ΔpH), K^+ uniport ($\Delta\psi$), and $2H^+$. glutamate$^-$ symport and H^+. lactose symport (Δp). Similarly, the efflux of a metabolic end-product may specifically generate one or both components of Δp at the expense of the solute gradient. Thus, unlike ATP synthesis and hydrolysis, which are independent of the composition of Δp, the transport of a particular solute may be uniquely dependent upon, or may uniquely generate, ΔpH and/or $\Delta\psi$.

If $\Delta\bar{\mu}_S$ is in thermodynamic equilibrium with Δp, it can be predicted that

$$\Delta\bar{\mu}_S = (m + n)\Delta\psi - n.Z\Delta pH$$

where m is the charge number on the solute and n is the $\rightarrow H^+$/solute quotient. Thus for K^+ uniport $\Delta\bar{\mu}_S = \Delta\psi$, whereas for H^+.lactose symport and $2H^+$.glutamate symport $\Delta\bar{\mu}_S$ is equal to $\Delta\psi$-$Z\Delta pH$ and $\Delta\psi$-$2Z\Delta pH$ respectively. However, recent work with whole cells of *E. coli* and *Staphylococcus aureus* shows that at least in these two cases $\Delta\bar{\mu}_S$ is significantly lower than the driving force, thus indicating that solute accumulation attains a kinetic steady state rather than a thermodynamic equilibrium. Indeed, there is increasing evidence that the size of the internal solute pool is largely governed by dissipative leakage pathways, which include the metabolism of the solute and/or its incorporation into cell material, as well as genuine efflux from the cell via carrier-mediated processes and/or simple diffusion.

Bacterial Respiration and Photosynthesis

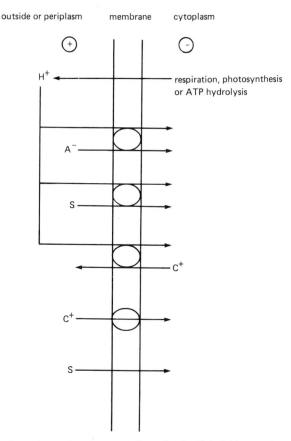

Fig. 5.6 Δp-dependent solute transport. Note that in *H halobium* and possibly also in some other organisms, $\Delta\mu_{Na^+}$ ($\equiv \Delta\psi + \Delta\mu_{Na^+}$) replaces Δp as the driving force for solute uptake.

Such leakage pathways probably serve to prevent the build-up of osmotically intolerable solute concentrations in response to particularly high driving forces.

A likely mechanism of Δp-dependent solute uptake is one in which the affinity and/or availability of the carrier for the solute are influenced by Δp, i.e. increased on the outer surface of the membrane and decreased on the inner surface. Thus it is envisaged that the carrier readily binds the external solute, then undergoes a small conformational change such that the binding site faces the cytoplasm, and finally releases the solute into the internal compartment. Several solute carriers have recently been purified and successfully reconstituted into proteoliposomes to exhibit uncoupler-sensitive transport.

Although active transport at the expense of $\Delta\bar{\mu}_{Na^+}$ is the norm in eukaryotes, it has so far been observed in relatively few species of bacteria, most importantly in some marine and alkaliphilic organisms and in the extreme halophile *H. halobium*. $\Delta\bar{\mu}_{Na^+}$

is generated in two ways in the latter organism, i.e. (i) indirectly via the light-driven bacteriorhodopsin-dependent H^+ pump and an electrogenic $H^+.Na^+$ antiport ($\rightarrow H^+/Na^+$ quotient > 1), and (ii) directly via the light-driven Na^+ pump; its main function is undoubtedly to drive the accumulation of amino acids.

The K^+ and Na^+ gradients ($[K^+]_{in} > [K^+]_{out}$ and $[Na^+]_{in} < [Na^+]_{out}$) generated at the expense of $\Delta\psi$ and ΔpH by K^+ uniport and $H^+.Na^+$ antiport systems serve three major functions. They (i) act as buffers to stabilize Δp, (ii) provide energy stores of considerably larger capacity and duration than are possible using H^+ gradients, (the latter being limited both by the necessity of maintaining an approximately neutral internal pH and by the relatively low H^+ buffering power of the cytoplasm), and (iii) satisfy the distinct preference of many cytoplasmic enzymes for K^+ rather than Na^+. It is not known which of these properties provided the major pressure for the evolution of these transport systems.

Cell motility Many species of bacteria exhibit motility, i.e. smooth forward runs alternating with relatively brief head-over-tail changes in direction (tumbling). According to species, motility is executed via one or more semi-rigid flagella and is controlled by a variety of external chemical stimuli (chemotaxis). Analyses of the effects of respiratory chain and ATP phosphohydrolase inhibitors, uncoupling agents, Unc^- mutations and artificially-imposed ΔpH and/or $\Delta\psi$ on motility indicate that forward movement occurs at the expense of Δp provided that the latter exceeds a threshold value of approximately 90 mV (inside alkaline and electrically negative); below this value, forward movement declines and the frequency of tumbling is greatly increased.

The mechanism by which Δp drives flagellar rotation is far from clear. Each flagellum exits through a basal hook which is embedded in the cell envelope via a rod which passes through up to four rings (L, P, S and M; the two outermost rings, L and P, are absent from Gram positive bacteria). The rod is firmly attached to the M ring in the coupling membrane, such that rotation of this ring by some form of protonic motor driven by Δp causes rotation of the flagellum; anticlockwise rotation is responsible for forward movement whereas sudden reversal leads to tumbling. The proportion of the total time given to these two types of movement determines the progress of the organism in a given direction and forms the basis of the chemotactic response. Thus attractants (e.g. D-galactose) and repellents (e.g. acetate) respectively decrease and increase the frequency of tumbling compared with that observed in the presence of chemotactically inert solutes (e.g. glycerol). There is some evidence that attractants cause demethylation of specific 'chemotaxis proteins' and in so doing may impose a particular ion specificity on a gated channel in the M ring, such that the resultant Δp-driven ion flux predominantly favours anticlockwise rotation (and hence forward movement up the attractant gradient); in contrast, repellents increase methylation and thus may impose an alternative ion specificity such that clockwise rotation of the flagella is preferred (and hence increased tumbling down the repellent gradient).

The gliding of cyanobacteria is also powered by a Δp-dependent motor, in this case causing rotation of fibrils located between its outer membrane and the peptidoglycan layer such as to induce a helical wave motion in the cell wall.

Bacterial Respiration and Photosynthesis

Summary

The energized coupling membrane generated by respiration or photosynthesis catalyses various energy-dependent reactions including ATP (or pyrophosphate) synthesis, reversed electron transfer, certain forms of solute transport and cell motility; in fermentative anaerobes, membrane energization is effected by hydrolysis of the ATP produced via substrate level phosphorylation.

The ATP phosphohydrolase ($BF_0.BF_1$) is a large, membrane-bound enzyme complex which is composed of 4 to 8 different subunits according to species. The smaller enzyme principally catalyses ATP hydrolysis, whereas the major function of the larger form is to catalyse ATP synthesis. Both reactions are associated with proton movement which, at least *in vitro*, can lead to the dissipation or generation of a transmembrane Δp; either component of the latter ($\Delta\psi$ or ΔpH) will drive ATP synthesis provided that the driving force exceeds a threshold value. Most determinations of the $\rightarrow H^+/ATP$ quotient indicate a value of approximately 2 to 3 g-ion H^+.mole ATP^{-1}; it is possible that this value may be species-dependent. ATP synthesis is probably effected via an alternating catalytic site mechanism, energized protons being required to facilitate the binding of the reactants and the release of the products, rather than for the condensation of ADP and phosphate *per se*.

Unlike ATP synthesis, for which the composition of Δp is unimportant, some types of solute accumulation specifically require $\Delta\psi$ and/or ΔpH according to the nature of the transported species. Similarly, the exit of metabolic end-products down their own concentration gradients can specifically generate $\Delta\psi$ and/or ΔpH, which in turn can drive ATP synthesis (thus increasing the ATP yield during fermentative growth). In some organisms, including the extreme halophile *H. halobium*, the accumulation of certain solutes is driven by Δp and/or $\Delta\bar{\mu}_{Na}^+$.

Reversed electron transfer, pyrophosphate synthesis and cell motility also occur at the initial expense of energized protons, but whether the latter constitute a localized or delocalized energy current remains to be elucidated.

References

BAIRD, B. A. and HAMMES, G. G. (1979). Structure of oxidative and photophosphorylation coupling factor complexes. *Biochimica et biophysica Acta* 549: 31–53.

DOWNIE, J. A., GIBSON, F. and COX, G. B. (1979). Membrane adenosine triphosphatases of prokaryotic cells. *Annual Review of Biochemistry* 48: 103–31.

FILLINGAME, R. H. (1980). The proton-translocating pumps of oxidative phosphorylation. *Annual Review of Biochemistry* 49: 1079–1113.

HAROLD, F. M. (1979). Vectorial metabolism. In: *The Bacteria* Vol 6. pp. 463–521. Edited by L. N. Ornston and J. R. Sokatch. Academic Press, New York and London.

KAGAWA, Y. (1978). Reconstitution of the energy transformer, gate and channel, subunit reassembly, crystalline ATPase and ATP synthesis. *Biochimica et biophysica Acta* 505: 45–93.

KELL, D. B. (1979). On the functional proton current pathway of electron transport phosphorylation; an electrodic view. *Boichemica et biophyscia Acta* 549: 55–99.

SIMONI, R. D. and POSTMA, P. W. (1975). The energetics of bacterial active transport. *Annual Review of Biochemistry* 44: 523–54.

Index

acidophiles 11, 35, 36, 38, 59
adenine nucleotide translocase 60
adenosine-5'-diphosphate (ADP) 1, 8
adenosine-5'-triphosphate (ATP) 1, 8
Alcaligenes (*Hydrogenomas*) 18, 20, 27, 28, 32, 39, 40, 56
alkaliphiles 11, 35, 36, 99
adenosine-5'-phosphosulphate (APS) 52, 53
APS reductase 52, 53, 54
Archaebacteria 59
Arthrobacter 20
ATP phosphohydrolase (ATPase-ATP synthetase) 2, 6, 86–96
   BF_0 6, 9, 12, 86–96
   BF_1 6, 9, 12, 86–96
   mutants 91–94
ATP/O (ATP/$2e^-$)quotient 6, 30, 31, 43, 45, 47, 59, 60, 81
ATP sulphurylase 52, 54
Azotobacter 20, 23, 24, 25, 31, 32, 49, 56
azurin 49
Bacillus 20, 27, 29, 31, 32, 35
bacteriochlorophyll 4, 64–73
bacteriopheophytin 4, 64–72
bacteriorhodopsin 6, 8, 81–85
bisulphite reductase 54
blue-green bacteria 4, 5, 10, 66, 67, 78–81
Bohr effects 8, 31
carbon dioxide 3, 4, 5, 38, 59
carbon monoxide 4, 17, 38, 61
carotenoids 4, 64–72, 75, 76, 82
catalase 17
cell motility 2, 103
→charge(K^+)/O quotients 27, 29
chemiosmotic hypothesis 6–10, 35, 36, 98, 99
chemoheterotrophs 2, 10, 14–37
chemolithotrophs 2, 3, 38–63
chemotaxis 103
Chlorobium 52, 66, 67
Chloroflexus 52, 66, 67
chlorophyll 4, 78–81
chlorosomes 10
Chromatium 49, 52, 66–72, 86
Chromobacterium 21
Clostridium 49, 86–90
coenzyme M 60
colicins 24
conformational changes 9, 98
copper proteins 3, 17, 19, 49, 57
cytochromes 2, 3, 16–25
   a-type 16–19, 28
   b-type 16–18, 21, 24
   c-type 16–22, 24, 28, 32, 35, 50, 55, 72, 75
   d-type 16–21, 28
   P460 43
cytochrome oxidases 2, 16–19, 24, 32
dehydrogenases 2, 5, 14, 15
   formate 46, 47
   lactate 15, 25, 54
   malate 25
   methanol 15, 28, 29
   NADH 15, 18, 21, 28

   NADPH 24
   succinate 15, 18, 25
Desulfotomaculum 40, 51, 53–55
Desulfovibrio 40, 51, 53–56
Desulfuromonas 51, 54
Energy coupling sites 28–32
Escherichia 15, 17–28, 31–33, 40, 45, 49, 56–61, 86–94, 101
F_{420} 60
fermentation 2, 3, 15
ferredoxin 54, 55
ferric iron (Fe^{3+}) 3, 38, 58
ferrous iron (Fe^{2+}) 3, 38, 57
flavin adenine dinucleotide (FAD) 14, 15
flavin mononucleotide (FMN) 14, 15
flavin oxidase 17
flavodoxin 50
flavoproteins 2, 3, 14
fumarate reductase 58, 59
Gram stain 10–12
green bacteria 4, 5, 10, 64–78
Haemophilus 15
→H^+/ATP quotient 8, 9, 96–98
Halobacterium 66, 67, 81–84, 103
halophiles 11
→H^+/O(→H^+/$2e^-$)quotient 7, 8, 9, 27–30, 35, 43–49, 75–77
hydrogenase 54–56
hydroxylamine-cytochrome *c* reductase 43
Hyphomicrobium 15
inhibitors
   of ATP phosphohydrolase 86, 89, 93
   of respiratory chain 18–21, 24, 49
ionophorous antibiotics 26, 33, 34, 76, 77, 95, 96
iron cycle 38, 57, 58
iron-sulphur proteins 2–4, 14, 15, 19, 44, 50
Klebsiella 20, 27, 40, 45, 49
Lactobacillaceae 17
light-harvesting complex 64–73
localized proton hypothesis 9, 10, 35, 98, 99
manganese proteins 5, 79
membranes 9–12
   cytoplasmic (energy coupling) 10–12, 18
   outer 10
membrane vesicles
   inside out 12, 25, 26, 30
   right-side-out 12, 25, 26, 30
Methanobacterium 59–61
methanogenesis 59–61
methanopterin 60
methoxatin (*see* pyrroloquinoline quinone)
Methylophilus 15, 20, 27, 28, 97
Micrococcus 20, 23, 24, 25, 27, 31, 32, 86
molar growth yields 30, 32, 43, 45, 49, 53, 57
molybdoproteins 3, 47
Mycobacterium 31, 32, 86
nicotinamide adenine dinucleotide (NAD^+, NADH) 2, 4, 5, 18, 25, 32
nicotinamide adenine dinucleotide phosphate ($NADP^+$, NADPH) 4, 5, 18, 25, 32
nitrate reductase 45–47
nitrite oxidase 44, 45, 48